Instructor's Manual for Jay's Modern Food Microbiology

Sixth Edition

Peter S. Murano, PhD
Assistant Professor
Department of Animal Science
Texas A & M University
College Station, Texas

Elsa A. Murano, PhD
Associate Professor
Department of Animal Science
Texas A & M University
College Station, Texas

AN ASPEN PUBLICATION®
Aspen Publishers, Inc.
Gaithersburg, Maryland
2000

Orders: (800) 638-8437
Customer Service: (800) 234-1660

Editorial Services: Joan Sesma
ISBN: 0-8342-1820-8

1 2 3 4 5

Table of Contents

Introduction v

Chapter 1 — History of Microorganisms in Food 1

Chapter 2 — Taxonomy, Role, and Significance of Microorganisms in Foods 3

Chapter 3 — Intrinsic and Extrinsic Parameters of Foods That Affect Microbial Growth 6

Chapter 4 — Fresh Meats and Poultry 8

Chapter 5 — Processed Meats 11

Chapter 6 — Seafoods 13

Chapter 7 — Fermentation and Fermented Dairy Products 15

Chapter 8 — Fruit and Vegetable Products: Whole, Fresh-Cut, and Fermented 18

Chapter 9 — Miscellaneous Food Products 21

Chapter 10—Culture, Microscopic, and Sampling Methods 24

Chapter 11—Physical, Chemical, Molecular, and Immunological Methods 27

Chapter 12—Bioassay and Related Methods 31

Chapter 13—Food Preservation with Chemicals 33

Chapter 14—Food Preservation with Modified Atmospheres 36

Chapter 15—Radiation Preservation of Foods and Nature of Microbial Radiation Resistance 38

Chapter 16—Low-Temperature Food Preservation and Characteristics of Psychrotrophic Microorganisms 41

Chapter 17—High-Temperature Food Preservation and Characteristics of Thermophilic Microorganisms **43**

Chapter 18—Preservation of Foods by Drying **47**

Chapter 19—Other Food Preservation Methods **50**

Chapter 20—Indicators of Food Microbial Quality and Safety **52**

Chapter 21—The HACCP System and Food Safety **55**

Chapter 22—Introduction to Foodborne Pathogens **58**

Chapter 23—Staphylococcal Gastroenteritis **62**

Chapter 24—Food Poisoning Caused by Gram-positive Sporeforming Bacteria **65**

Chapter 25—Foodborne Listeriosis **68**

Chapter 26—Foodborne Gastroenteritis Caused by *Salmonella* and *Shigella* **72**

Chapter 27—Foodborne Gastroenteritis Caused by *Escherichia coli* **74**

Chapter 28—Foodborne Gastroenteritis Caused by *Vibrio, Yersinia*, and *Campylobacter* Species **77**

Chapter 29—Foodborne Animal Parasites **80**

Chapter 30—Mycotoxins **84**

Chapter 31—Viruses and Some Other Proven and Suspected Foodborne Biohazards **87**

Introduction

This Manual is intended to be used as companion material to the textbook *Modern Food Microbiology*, Sixth Edition, by James Jay. Each chapter of the Manual corresponds directly to the chapters in the textbook, and is divided into the following sections:

1. Learning Objectives.
2. Chapter Outline.
3. Terms and Definitions.
4. Basic Knowledge Questions.
5. Critical Thinking Questions.

The purpose of the **Learning Objectives** section is to help the instructor design the lecture material in a way that will ensure a successful outcome, from the perspective of both the instructor and the student. There are two purposes for the **Chapter Outline**. One is to help the instructor organize the material for the lecture, aiding in the decision-making regarding which chapters and/or sections to include or omit. The second is to serve as a summary of the material that is included in the textbook, enabling the instructor to review it at a glance.

The purpose of the **Terms and Definitions** section of each chapter is to serve as supplementary material, providing definitions of terms that are either only briefly defined, or not defined at all in the textbook. Thus, the terms and definitions provided can help clarify concepts covered in the textbook. This section can also serve as a source of questions for quizzes or exams. It can also serve as review material to be used in summary discussions.

Finally, the two sections on questions are designed to test student-learning of the fundamental concepts from each chapter. The success of accomplishing the learning objectives is directly related to student performance in answering these questions. In particular, the **Basic Knowledge Questions** section is designed to review the material covered in the chapter, and serve as a good indicator of a type of learning known as "lower-level cognition." The **Critical Thinking Questions** section is designed to challenge critical thinking skills by providing students with questions that require a more complex type of learning known as "higher-level cognition." In some instances there is no "correct" answer to these questions, but the student is encouraged to develop a reasonable and defensible answer or hypothesis to explain phenomena for which no definitive answer has been provided by current research. The instructor is encouraged to use the questions provided in both of these sections in quizzes or exams, or to utilize them to launch discussions on specific topic areas in more advanced lecture settings, such as in honors-level or graduate-level courses.

CHAPTER 1

History of Microorganisms in Foods

Learning Objectives

The reader will be able to:

1. Discuss the events of ancient history related to the presence and use of microorganisms in food products.
2. Describe the contributions of notable microbiologists to our understanding of the role of microorganisms in food spoilage and food production.
3. List major historical events related to food law, food processing accomplishments, and outbreaks of foodborne illness that have shaped the science of food microbiology in the last 50 years.

Chapter Outline

1. Historical account of food microbiology
   a) Food preservation methods in ancient times
   b) The role of microorganisms in food poisoning and food spoilage
   c) Modern food processing
2. Significant dates and events
   a) Food preservation
   b) Food spoilage
   c) Food poisoning
   d) Food legislation

Terms and Definitions

Aflatoxins: Toxins produced by the mold genus *Aspergillus*.

BSE: Condition of cattle, also known as "mad cow disease", which manifests as tremors and lack of coordination in the animal, resulting in death due to brain damage.

Botulism: Disease caused by ingestion of toxin produced by the bacterium *Clostridium botulinum*, often resulting in death by paralysis of involuntary muscles.

Brine: Solution of water and salt, used to pack foods into cans to promote even heat distribution during heat processing.

Cyclosporiasis: Condition caused by infection with the parasite *Cyclospora cayetanensis*, usually from consumption of contaminated fresh foods such as fruits and vegetables, or water.

Ergot: Type of food poisoning caused by ingestion of toxin produced by the mold *Claviceps purpurea* when growing on wheat.

Gastroenteritis: Condition marked by inflammation of the intestinal tract, usually as a result of infection by microbial pathogens.

Giardiasis: Condition caused by infection with the protozoan parasite *Giardia lamblia*.

Hemorragic Colitis: Condition caused by infection with verotoxin-producing strains of *Escherichia coli*, most notably serotype O157:H7, manifesting itself as profusely bloody diarrhea.

Irradiation: Processing treatment based on exposure of foods to ionizing sources of radiation, such as gamma rays, accelerated electrons, or x-rays, for the purpose of reducing or eliminating microbial, parasitic, and insect contaminants.

Listeriosis: Condition caused by infection with the bacterium *Listeria monocytogenes*, which manifests itself as flu-like symptoms in normal, healthy adults, but develops into complications such as stillbirths, meningitis, and encephalitis in people with depressed immune systems.

Smoking (e.g. meats): A processing treatment whereby foods are exposed to smoke produced by burning of wood or other fuels, or by addition of extracted smoke products, for the purpose of changing the flavor and extending the shelf-life of the products.

Spontaneous Generation: The sudden appearance of living organisms or biological contaminants from no apparent source, a rejected theory once used by scientists to explain how diseases and the spoilage of food occurs.

Basic Knowledge Questions

1. What experiment did Spallanzani conduct to disprove the doctrine of spontaneous generation?
2. What processing treatment did N. Appert invent?
3. What two contributions did Pasteur make to the field of microbiology?
4. Who commercialized the freezing of foods in the U.S.?
5. What important food law was enacted in 1906 in the U.S.?
6. Name three microorganisms known to be responsible for outbreaks of foodborne illness only in the last 20 years.
7. What is the Talmadge-Aiken Act?

Critical Thinking Questions

1. Design an experiment that would show that ozone application to foods can kill spoilage bacteria.
2. How would our lives be different if we did not have canning as a food preservation method?

CHAPTER 2

Taxonomy, Role, and Significance of Microorganisms in Foods

Learning Objectives

The reader will be able to:

1. Discuss the various molecular methods used to classify bacteria.
2. List and describe the environments where microorganisms that contaminate food originate.
3. List and describe several types or groups of foodborne bacteria.
4. Describe the general physiology of molds.
5. List and describe several types of foodborne molds.
6. Describe the general physiology of yeasts.
7. List and describe several types of foodborne yeasts.
8. List ways in which bacteria, yeasts, and molds differ from each other.

Chapter Outline

1. Introduction to the presence and purpose of microorganisms.
2. Bacterial taxonomy.
   a) rRNA analyses
   b) Analysis of DNA
3. Primary sources of microorganisms in foods.
   a) Soil and water
   b) Plants and plant products
   c) Food utensils
   d) Gastrointestinal tract
   e) Food handlers
   f) Animal feeds
   g) Animal hides
   h) Air and dust
4. Synopsis of common foodborne bacteria.
5. Synopsis of common genera of foodborne molds.
   a) Basic mold physiology
   b) Common genera found in foods
6. Synopsis of common genera of foodborne yeasts.
   a) Basic yeast physiology
   b) Common genera found in foods

Terms and Definitions

Aerobe: An organism that requires oxygen for optimum growth. If a "strict aerobe", presence of oxygen is required for growth, e.g. *Bacillus* species.

Anaerobe: An organism that requires the absence of oxygen for optimum growth. If a "strict anaerobe", absence of oxygen is required for growth, e.g. *Clostridium* species.

Ascospore: Sexual spores formed within an enclosed sac, or ascus, of certain molds.

Cleistothecia: Carp in which ascospores develop within a closed envelope in certain molds.

Conidia: Asexual spores produced by mycelium of certain molds, used in reproduction. Conidia are resistant to UV radiation and to dry conditions.

Columella: The neck or tapering base of the sporangium of molds.

DNA-DNA Hybridization: The joining (annealing) of two single strands of DNA into one double strand, guided by the pairing of bases from one strand to the compatible bases in the other strand (adenine with thymine, and cytosine with guanine). If two organisms are related, their base sequences will be very similar, and thus DNA from one will anneal to that of the other.

Genus: Classification level of organisms, more specific than "family", but less than "species", based on broad physical, biochemical, genetic and/or other characteristics (plural is "genera").

Gram-negative: Bacteria that stain red as a result of gram-staining, due to their thin cell wall structure.

Gram-positive: Bacteria that stain blue/purple as a result of gram-staining, due to their thick cell wall structure.

Hetero-fermentative: An organism that generates various organic compounds as a result of carrying out fermentation reactions.

Hyphae: Filamentous structures that make up the mycelia of molds.

Koji: Cooked rice inoculated with *Aspergillus oryzae*, *A. soyae*, or *A. japonicus*, and allowed to ferment to produce soy sauce.

Mesophile: An organism that grows best at temperatures in the range of 20 to 40°C.

Mesotroph: An organism that can grow at temperatures between 20 and 40°C.

Microaerophile: An organism that grows best in a low-oxygen atmosphere. Often, the term is used to refer to organisms that cannot tolerate atmospheric levels of oxygen.

Microbiota: Microscopic environment, including all of its resident organisms.

Mol G + C: The percent of guanine and cytosine bases that contribute to the total base content in a DNA molecule. The DNA of most gram-negative bacteria has a G + C content > 55%, while that of most gram-positives has a G + C content < 50%.

Mycelium: Tangled mass of filaments, or hyphae, formed by the growth of some foodborne molds, visible to the naked eye.

Psychrotroph: An organism that can grow at or below 4°C and also at temperatures below 20°C.

Reverse Transcriptase: Enzyme that catalyzes the synthesis of DNA from RNA. It is used to build genetic libraries of DNA molecules from the rRNA of different organisms, with which studies can be carried out to determine their degree of relatedness.

Rhizoid: Root-like filament that functions as absorption organs of filamentous fungi.

Ribosomes: Cellular structures, composed of protein and rRNA (ribosomal RNA).

RNA: Ribonucleic acid, molecule found in living organisms, responsible for translating the encoded messages in genetic material into proteins.

Sanger method: Biochemically-based method used to determine the order, or sequence, of bases (adenine, cytosine, thymidine, and guanine) present in a piece of DNA.

Species: Classification of organisms within a specific genus, based on unique physical, biochemical, genetic and/or other characteristics.

Stolon: The prolonged extension of the body wall of some molds, from which buds develop.

Taxonomy: Classification of living things according to specific characteristics into distinct groupings.

Thermophile: An organism that generally requires temperatures of 55°C or above for growth.

Xerophile: An organism that grows best in low-humidity conditions.

Basic Knowledge Questions

1. Why is rRNA used in taxonomical classification of microorganisms?
2. Describe three sources of contamination of foods of animal origin.
3. What are some of the new genera into which organisms formerly classified as *Micrococcus* have been placed?
4. Describe the general structure of molds.
5. What is a teleomorph?
6. Name two genera of molds that are commonly responsible for spoilage of fruits.
7. Name and describe the genus of yeast used in the production of breads and beer.

Critical Thinking Questions

1. Describe the procedure you would follow to determine the source of contamination of a particular food.
2. Should molecular-based methods be used as the definitive way to classify microorganisms? Why or why not?

CHAPTER 3

Intrinsic and Extrinsic Parameters of Foods That Affect Microbial Growth

Learning Objectives

The reader will be able to:

1. Define pH, water activity, and O/R potential. Describe how these and other intrinsic parameters affect microbial growth.
2. Describe how various extrinsic parameters affect the ability of microorganisms to grow in foods.
3. Describe the concept of microbial interference and antagonism between microorganisms.
4. Explain the hurdle concept and describe examples of how it can be applied to inhibit and/or eliminate microorganisms in foods.

Chapter Outline

1. Intrinsic parameters.
   a) pH
      1. General facts
      2. Effects on microorganisms
   b) Moisture content
      1. Water activity (a_w)
      2. Effects of low a_w
   c) Oxidation-reduction potential
      1. General facts
      2. Eh effects
   d) Nutrient content
   e) Antimicrobial constituents
      1. General facts
      2. Lactoperoxidase system
   f) Biological structures
2. Extrinsic Parameters
   a) Temperature
   b) Relative humidity
   c) Concentration of gases
   d) Presence of other microorganisms
      1. General microbial interference
      2. Lactic antagonism
3. Combined Intrinsic Extrinsic Parameters: The Hurdle Concept

Terms and Definitions

Adsorption: The process of acquiring moisture by a system, typically occurring as a result of an excess of moisture in the environment surrounding that system.

Arrhenius Equation: Mathematical equation that explains the rate of reactions, such as the chemical reaction between an acid and a base to form a salt and water. In this equation, the rate constant, k, is directly and linearly related to the temperature, T, of the components.

Biological Acidity: Acidity of a food, caused by the action of microorganisms, typically through fermentation reactions.

β-galactosidase: Enzyme that catalyzes the breakdown of galactose by microbial cells.

Buffer: A solution or food component that, because of its biochemical structure, resists changes in the pH of its surroundings.

Catabolism: Process of breaking down biochemical compounds into usable molecules by a biological system.

Desorption: The process of losing moisture by a system, typically occurring as a result of a deficit of moisture in the environment surrounding that system.

Generation Time: Time required for cell division to take place. A generation time of 20 min. refers to the time it takes for one cell to divide, yielding two cells.

Lag Phase: Phase of bacterial growth in which synthesis of macromolecules is being carried out, such as for repairing of damaged structures, and for acclimatization to new environmental conditions. In this phase, cell division does not take place.

Osmotic Stress: Condition in which a microorganism is exposed to a high concentration of solutes in the surrounding environment, causing diffusion of water from the inside to the outside of the cell by osmosis, in order to equalize the pressure on both sides.

Oxidation-Reduction Potential: The ease with which a particular substrate gains or loses electrons. A substrate that loses electrons (becomes oxidized), and causes other substrates to gain electrons (become reduced). A system with a high O/R potential lacks electrons, thus it is oxidized. Conversely, a system with a low O/R potential has an oversupply of electrons, thus it is reduced.

Poising Capacity: The tendency of a system to resist changes in O/R potential.

Shelf-life: Period of time required for a food to become spoiled and thus organoleptically undesirable.

Water Activity: The amount of available water in a medium for microbial growth. It is also defined as the ratio of vapor pressure of a food to the vapor pressure of pure water.

Basic Knowledge Questions

1. Explain how weak acids (i.e. organic acids) affect bacterial cells.
2. Provide the mathematical definition of water activity (a_w), and explain how having a low a_w in the medium inhibits cell multiplication.
3. Name four factors that determine the O/R potential of a food system.
4. Name five types of nutrients required by microorganisms.
5. How does the lactoperoxidase system inhibit the growth of microorganisms?
6. Name four extrinsic parameters that affect microbial growth.
7. Define "psychrotroph" and "thermophile."
8. Define "bacterial interference" and give two examples of the types of interference that can be observed between two organisms.
9. Give an example of how the hurdle technology can be applied to prevent the growth of microorganisms.

Critical Thinking Questions

1. A new microorganism has been discovered, and it has been reported that it is able to grow only in the presence of bacteria used in the fermentation of milk. Provide a possible scenario or mechanism to explain this phenomenon.
2. What is the relationship between O/R potential and the oxygen requirements of microorganisms (i.e. aerobes, anaerobes, etc.)?

CHAPTER 4

Fresh Meats and Poultry

Learning Objectives

The reader will be able to:

1. List the sources of contamination of fresh meats.
2. List the steps involved in the slaughter of food animals.
3. Explain the reasons why ground meats are more highly contaminated than whole cuts.
4. Describe the differences between various types of meats (i.e. hot-boned vs. cold-boned meats, variety meats, etc.) in terms of numbers and types of microorganisms present.
5. Describe the types of organisms responsible for spoilage of fresh red meats.
6. List examples of methods that can be used to detect microbial spoilage of meats.
7. Describe the events that take place during spoilage of liver.
8. Describe the spoilage of whole, cut-up, and comminuted poultry.
9. List and describe carcass washing and sanitation procedures employed by the industry to decrease microbial contaminants.

Chapter Outline

1. Sources of contamination
2. Biochemical events that lead to *rigor mortis.*
3. The biota of meats and poultry.
4. Incidence/prevalence of microorganisms in fresh red meats.
   a) Traditional ground meats
   b) Soy-extended ground meats
   c) Mechanically deboned meat, poultry, and fish
   d) Hot-boned meats
   e) Effect of electrical stimulation
   f) Organ and variety meats
5. Microbial spoilage of fresh red meats.
   a) General considerations
   b) Mechanism of spoilage
6. Spoilage of fresh livers.
7. Incidence/prevalence of microorganisms in fresh poultry.
8. Carcass sanitizing/washing.

Terms and Definitions

Actin: Protein found in thin filaments of muscle tissue cells. It binds to another protein, myosin, resulting in muscle contraction.

Bone Taint: Spoilage of fresh meat near the bone, typically caused by *Clostridium* and *Enterococcus* species.

Cathepsins: Enzymes that break down proteins in tissue cells, resulting in the release of amino acids.

Coliform: A gram-negative, non-spore-forming rod-shaped bacterium, able to ferment lactose with gas production within 48 hrs. at 35°C.

DFD: Dark, firm, and dry meat, it typically has a pH higher than that of normal meat tissue, causing it to spoil faster.

Fecal Coliform: A gram-negative, non-spore-forming rod-shaped bacterium, able to ferment lactose with gas production within 48 hrs. at 44.5°C. Since most microorganisms that are shed in the feces of humans and animals share these characteristics, fecal coliforms are used as indicators of the presence of fecal contamination in foods.

Lymph Nodes: Organs that form part of a network which filters foreign bodies from the tissue fluid and lymph, collecting microbial and other particulate contaminants. Cells of the immune system called lymphocytes reside in the nodes, and play a role in digestion and/or inactivation of invading organisms.

MPN: Most probable number method, used to estimate the number of cells of a particular class of organisms in a food sample, when that number is expected to be too low to be detectable by direct plating methods.

Myofibril: Structure inside a multinucleated skeletal muscle cell (fiber), composed of thick and thin filaments.

Myosin: Protein found in the thick filaments of myofibrils. It binds to another protein, actin, and catalyzes the breakdown of ATP, which drives muscle contraction.

PCR: Polymerase chain reaction, is a molecular biology-based method used to increase the amount of DNA of a particular organism in a sample, thereby increasing its detectability by subsequent assays.

Putrefactive Anaerobe: Literally, an organism that breaks down protein and grows optimally in the absence of oxygen. An example is *Clostridium sporogenes*, a bacterium that is used often as an indicator of commercial sterility.

Reticulo-endothelial System: Portion of the immune system of humans and animals that is responsible for the production of phagocytic cells (i.e. macrophages, polymorphonuclear leukocytes, etc.). These cells scavenge the body through the circulatory and lymphatic systems in order to remove invading particulates, including microorganisms.

Rigor mortis: Literally "stiffness of death," caused by depletion of ATP, which results in tension by the interaction of actin and myosin in muscle tissue.

Rumen: The first of four chambers, or stomachs, of a ruminant animal (i.e. cattle). Upon consumption of plants by the animal, the unchewed material first passes into the rumen. It is then regurgitated and masticated with saliva in the mouth. The material is then re-swallowed. It passes through the rumen, as well as two other chambers (the reticulum and the omasum) on its way to the third chamber (the abomasum), where it is digested.

Sarcoplasm: Fluid surrounding contractile myofibrils in a muscle cell. The term "sarcoplasmic fraction" refers to water-soluble proteins, obtained by extraction of the sarcoplasmic fluid.

Serovar: Classification within a microbial species, based on antigenic differences (i.e. type of antibodies that are produced as a result of exposure by a human or animal to that

	particular organism). For example, *Escherichia coli* O157:H7 and *Escherichia coli* O104:NM are two serovars (also termed "serotypes") of the same organism.
Visceral Taint:	Microbial spoilage of carcasses in New York-dressed poultry, due to invasion of inner tissues by microorganisms in the intestinal tract of the animal.

Basic Knowledge Questions

1. Describe how ground meat products become contaminated.
2. List several types of bacteria and molds that are responsible for spoilage of fresh meat and poultry.
3. Name at least three methods that can be used for detecting spoilage of fresh meats for each of the following categories:
   a) chemical methods
   b) physical methods
   c) bacteriological methods
   d) physicochemical methods
4. Describe how fresh meat becomes spoiled, both in the presence and absence of oxygen. Include the names of the types of organisms responsible.
5. What are the main differences between spoilage of fresh meats vs. organ (variety) meats?
6. Name two products of putrefaction, and describe the chemical reactions by which they are formed.
7. What are some of the decontamination methods used to remove microorganisms from animal carcasses?

Critical Thinking Questions

1. Compare and contrast the role of molds vs. bacteria in spoilage of fresh meat and poultry.
2. Rank the following products according to how quickly you would expect them to spoil, and justify your reasoning:
   a) chicken cuts (with skin)
   b) beef cuts
   c) vacuum-packaged ground beef

CHAPTER 5

Processed Meats

Learning Objectives

The reader will be able to:

1. Define the curing and smoking processes used in the meat industry.
2. Describe the biota of cured meats in terms of microbial composition and typical level of each contaminant.
3. Describe the types of spoilage that occur in cured and smoked products.
4. Describe the various types of fermented meat products.
5. Discuss some of the safety concerns associated with fermented meat products.

Chapter Outline

1. Common processing treatments.
   a) Curing
   b) Smoking
2. Sausage, bacon, bologna, and related products.
   a) Processing methods
   b) Spoilage
3. Bacon and cured ham.
4. Fermented meat products.
   a) Processing methods
   b) Product safety

Terms and Definitions

Choleglobin:	Green compound produced by oxidation of myoglobin or oxymyoglobin in meats, often as a result of microbial spoilage.
Dry Sausage:	Sausage product containing 30–40% moisture, not smoked or subjected to heat-processing.
Greening:	Condition of spoiled cured meat, caused by (1) oxidization of porphyrin molecule in curing agents as a result of reaction with hydrogen peroxide, the latter being produced by certain microorganisms, or by (2) production of hydrogen sulfide by microorganisms, which reacts with myoglobin to form the green-colored sulphmyoglobin.
Hetero-fermenters:	Microorganisms that ferment sugars to produce various compounds, such as organic acids, diacetyl, carbon dioxide, etc.
Homo-fermenters:	Microorganisms that ferment sugars to produce one compound, typically lactic acid or other organic acid.

Nitrite:	Curing agent used in processing some meat products, for the purpose of fixing tissue pigment, retain moisture, and prevent germination of *Clostridium botulinum* spores.
Offal:	Parts of an animal that are removed during dressing, especially but not limited to, digestive tract, lungs, blood, feet, and head.
Semidry Sausage:	Sausage products containing 50% moisture, heat-treated to achieve 60–68°C internal temperature, and smoked.
Sliminess:	Condition of spoiled cured meat, caused by production of long-chain carbohydrates by certain foodborne bacteria and yeasts, which impart a slimy texture to the product.
Souring:	Condition of spoiled cured meat, caused by production of organic acids through the fermentation of lactose by some bacteria, such as lactobacilli and enterococci.
Sours:	Condition of spoiled cured hams, caused by microbial fermentation of sugars pumped into the product.
Trichinosis:	Health condition caused by infection with larvae of the worm *Trichinella spiralis*, typically found in untreated raw pork.

Basic Knowledge Questions

1. Describe the purpose for the following ingredients in curing meats:
   a) nitrite
   b) sodium ascorbate
   c) sugar
   d) phosphates
2. Name at least four reasons why meats are smoked.
3. Define the following terms related to spoilage, and give an example of a microorganism that may cause it:
   a) sliminess
   b) souring
   c) greening
4. Compare the spoilage that takes place in cured hams vs. smoked hams.
5. Compare dry and semi-dry sausages in terms of processing treatments and types of organisms used to produce them.
6. Name the types of molds used in production of processed meats.
7. What are some concerns associated with fermented meat products and the organism *Escherichia coli* O157:H7?

Critical Thinking Question

1. You are asked to develop a new meat product that is shelf-stable, yet contains no nitrite. Describe the process and ingredients you would use, and justify your choices.

CHAPTER 6

Seafoods

Learning Objectives

The reader will be able to:

1. Define what is meant by "seafoods."
2. Describe the sources of contamination of fresh seafood.
3. Name some of the organisms that make up the microflora of fresh seafood.
4. Describe the processes involved in production of several fermented fish products.
5. Describe the changes that occur during spoilage of fresh fish, including decarboxylation of specific amino acids by microorganisms.
6. Discuss the types of spoilage flora that exist in crustaceans and mollusks, and the changes that take place as a result.

Chapter Outline

1. Definition of seafood.
2. Microbiological quality of various fresh and frozen products.
   a) Typical organisms making up the seafood microflora
   b) Levels of various types of microorganisms found in seafood
3. Fermented fish products.
   a) Fish sauces
   b) Fish pastes
4. Spoilage of fish and shellfish.
   a) Fish
   b) Shellfish
      1) Crustaceans
      2) Mollusks

Terms and Definitions

Alpha-hemolysis: Type of hemolysis observed on blood agar plates as a result of microbial enzymatic action on red blood cells in the agar. It manifests itself by formation of a cloudy green-colored zone of hemolysis surrounding the colonies of specific types of microorganisms that may be growing on the surface of the agar. It is considered a mild reaction, when compared with beta-hemolysis of red blood cells (exemplified by formation of a completely clear zone around the colonies).

Blanching: Processing treatment designed to heat foods at temperatures near boiling for a few seconds. It is considered a mild treatment yet it can serve to partially decontaminate the surface of some products, as well as inactivate product enzymes.

Halophilic: Microorganisms that grow optimally in the presence of high salt concentrations.

Histamine: Volatile compound produced as a result of the break down of the amino acid histidine in fish tissue by the enzyme histidine decarboxylase, which is found in certain microorganisms. Production of histamine is associated with microbial spoilage of these products, as well as scombroid food poisoning.

Shucking: Process used to remove oysters from their shell.

TMAO: Trimethylamine-N-oxide, compound found in fresh fish which is reduced to trimethylamine (TMA) during microbial spoilage.

Tyramine: Produced from decarboxylation of the amino acid tyrosine found in fish tissue, and is thus associated with spoilage of fish products.

Basic Knowledge Questions

1. What are some genera of microorganisms typically found in fresh seafood?
2. How are fermented fishery products made? Give some examples.
3. Name the types of microorganisms responsible for spoilage of fresh fish.
4. What are the components of slime?
5. What are the chemical changes that take place during spoilage of fresh fish?
6. Define the following terms:
   a) TVB
   b) TVA
   c) TVS
   d) TVN
7. What is the pH scale on which spoilage of oysters is based?

Critical Thinking Questions

1. Describe the shifts that occur regarding the prevalence of specific populations of microorganisms during storage of fresh fish. Speculate on the reasons why these shifts occur.
2. What are the physiological differences between crustaceans and mollusks that would affect the types of microorganisms that would be responsible for spoilage of each.

CHAPTER 7

Fermentation and Fermented Dairy Products

Learning Objectives

The reader will be able to:

1. Define "fermentation."
2. List the genera of organisms encompassed in the "lactic acid bacteria" group.
3. Provide an overview of biochemical pathways involved in fermentation of glucose by microorganisms.
4. Define "homofermentative" vs. "heterofermentative" organisms, and give examples of each.
5. Describe how microbial growth is measured.
6. Describe the types of organisms found in raw milk.
7. Define "pasteurization" treatments.
8. Describe microbial spoilage of milk.
9. Discuss the types of starter cultures used in dairy fermentation and provide examples of fermented dairy products.
10. Describe the steps involved in fermentation of milk to produce cheese.
11. Define "lactose intolerance."
12. Discuss the cholesterol-lowering and anticancer effects of fermented products.
13. Define "probiotics."
14. Discuss the disease-causing ability of lactic acid bacteria.

Chapter Outline

1. Fermentation.
   a) Background
   b) Definitions and characteristics of fermentation
   c) The lactic acid bacteria
   d) Metabolic pathways and molar growth yields
2. Dairy products.
   a) Milk biota
   b) Starter cultures, products
   c) Cheeses
3. Apparent health benefits of fermented milks.
   a) General
   b) Lactose intolerance
   c) Cholesterol
   d) Anticancer effects
   e) Probiotics
4. Diseases caused by lactic acid bacteria.

Terms and Definitions

Carcinoma: Malignant growth originating in epithelial tissue.

Cloning Vector: Any small genetic element that can be used to carry other genes, helping to replicate them inside a host.

Cloning vectors are usually designed to allow integration of foreign DNA for the purpose of transporting that DNA into whatever organism the vector is to be placed.

Crohn's Disease: Inflammatory bowel disease affecting humans, and suspected of being caused by species of *Mycobacterium*, the agent responsible for a similar condition in cattle and sheep known as Johne's Disease.

Curd: Precipitated casein protein, as a result of a lowering of the pH during the fermentation of milk by lactic acid bacteria.

Hetero-fermentative: Refers to the fermentation that some organisms undergo, in which glucose is metabolized by a series of reactions, one of which includes the enzyme phosphoketolase, to produce lactic acid, ethanol, and other compounds such as acetic acid, and carbon dioxide.

Homo-fermentative: Refers to the fermentation that some organisms undergo, in which glucose is metabolized by a series of reactions, one of which includes the enzyme hexose isomerase, to produce lactic acid.

Johne's Disease: Inflammatory bowel disease of sheep and cattle, caused by infection with *Mycobacterium avium* subspecies *paratuberculosis*. Similar to Crohn's Disease in humans, for which a causative agent has not been identified.

Molar Yield Constant: Can be defined as grams of cells formed per mole of nutrient consumed ($Y_{x/m}$), or grams of cells formed per mole of oxygen consumed ($Y_{x/O2}$), or grams of cells formed per moles of ATP formed. The overall goal is to calculate the yield of cells according to some medium component that is either used or formed as a result of cell growth.

Mucopeptide: A class of glycoproteins in which peptides are complexed with an acid, such as muramic acid, as occurs in the cell wall structure of most bacteria.

Peptidoglycan: Polymer found in the cell wall of most bacteria, made up of N-acetylglucosamine and muramic acid molecules held together by amino acid bridges.

Vector Plasmid: Small, circular, double-stranded DNA that serves as a vehicle to insert specific sequences, or genes, into a particular organism. In the case of starter cultures, a vector plasmid is constructed with a gene that encodes for the production of specific compounds. This vector plasmid is then introduced into the particular microorganism that is to carry out the fermentation reactions.

Probiotic: A food that contains a microorganism, or consortium of microorganisms, ingestion of which is thought to promote health and well-being of the consumer.

Sarcoma: Tumor of fleshy consistency, a malignant growth derived from non-epithelial tissue (i.e. connective tissue, lymphoid tissue, cartilage, etc.).

Starter Culture: A culture of various types of microorganisms, typically lactic acid bacteria, which is used for fermentation of milk and other commodities for the production of fermented foods. These are often genetically-modified in order to promote production of specific compounds during fermentation, and to minimize variations in production yields.

Titratable Acidity: Used to monitor the progress of

fermentation by measuring the amount of base required to neutralize the amount of acid in a product, i.e. lactic acid. It provides a measure of the total acidity rather than of hydrogen ion concentration.

Basic Knowledge Questions

1. What is fermentation? Include in your answer a definition for "top" and "bottom" fermentations.
2. List the 12 genera of bacteria that constitute the "lactic acid bacteria," and describe them in terms of growth requirements and other general characteristics.
3. What is "molar growth yield"?
4. Describe the spoilage of pasteurized milk products.
5. Name some organisms that are commonly used as starter cultures in dairy fermentation.
6. Describe production of yogurt, including the organisms responsible and the interaction between the two that results in the final product.
7. What is lactose intolerance and how can it be treated?

Critical Thinking Questions

1. Compare and contrast homofermentation and heterofermentation. Include differences in terms of metabolic pathways, enzymes, and examples of microorganisms.
2. You work at a cheese production plant. A new strain of *Propionibacterium* has been purchased by the company and you are asked to determine whether it is as good, or better, than the strain that is currently being used in the production of swiss cheese. Design an experiment and the approach you would follow to answer this question.

CHAPTER 8

Fruit and Vegetable Products: Whole, Fresh-Cut, and Fermented

Learning Objectives

The reader will be able to:

1. Describe how fruits and vegetables become contaminated with microorganisms.
2. Describe the characteristics of vegetables that make them prone to specific types of spoilage.
3. Define the various types of spoilage conditions of vegetables caused by bacteria and by molds, and include examples of organisms responsible for these.
4. Describe the spoilage of fruit, including a discussion of the characteristics that enable these commodities to resist growth by most microorganisms.
5. Define fresh-cut produce and discuss the types of spoilage and disease-causing microorganisms that may be found.
6. Describe various plant-based fermented products, such as breads, olives, pickles, sauerkraut, wine, and cider within the context of the processes involved and the types of starter cultures utilized to produce them.
7. Describe fermented products made from beans of various plants, such as coffee and cacao.
8. Describe soy-based and other crop-based fermented products, including the process involved and the organisms used for fermentation reactions.

Chapter Outline

1. Fresh and frozen vegetables.
   a) Incidence of microorganisms
   b) Spoilage
      1) Bacterial agents
         a. Soft rot
         b. Other conditions
      2) Fungal agents
         a. Gray mold rot
         b. Sour rot
         c. Rhizhopus soft rot
         d. Phytophora rot
         e. Anthracnose
2. Spoilage of fruits.
3. Fresh-cut produce.
   a) Examples
   b) Microbial load
   c) Pathogens
4. Fermented products.
   a) Breads
   b) Olives, pickles, and sauerkraut
      1) Olives
      2) Pickles
      3) Sauerkraut
   c) Beer, ale, wines, cider, and distilled spirits
      1) Beer and ale
      2) Wines
      3) Cider

4) Distilled spirits
5) Miscellaneous fermented products
a) Coffee beans
b) Cacao beans
c) Soybean products
1) Soy sauce
2) Tempeh
3) Miso
d) Ogi
e) Gari
f) Bongkrek
g) Ontjom

Terms and Definitions

Auxins: Plant hormones that can desensitize plants to ethylene and thus delay ripening.

Blight: Any disease, symptom of disease, or injury of plants resulting in withering, cessation of growth, etc. but without rotting.

Catalase: Enzyme found in many microorganisms, which catalyzes a reaction that breaks down hydrogen peroxide to release oxygen and water. It is part of the defense mechanism of bacteria to prevent interference of hydrogen peroxide in its biochemical reactions.

Cellulose: Complex carbohydrate that forms the wall of plant cells, composed of up to 12,000 glucose molecules linked by beta 1–4 linkages.

Cotyledons: Storehouse of nourishment for the embryo of seed plants, the leaf in a seed.

Ethylene: Plant hormone known to trigger ripening of fruit and promote plant senescence.

Hops: Flowers used to impart certain flavors, aromas, and bitterness to wort during beer production.

Lenticel: One of the cortical pores in the stems of woody plants by which air penetrates.

Lye: Strongly alkaline solution such as NaOH.

OTR: Oxygen transmission rate, usually refers to the property of a packaging film to allow the free passage of oxygen. For example, a low OTR film is used for packaging products under vacuum.

Pathovar: Classification of microorganisms below the species level according to their ability to cause specific pathogenic effects.

Pectin: Complex carbohydrate found in the cell wall of most fruits and vegetables, composed of galacturonic acid molecules linked together by alpha 1–4 linkages.

Superoxide: Radical of oxygen containing an unpaired electron. Highly reactive molecule, capable of interfering with many biochemical reactions of bacteria and other organisms.

Superoxide Dismutase: Enzyme found in many microorganisms, which catalyzes a reaction that combines two superoxide molecules and water to form hydrogen peroxide and oxygen.

Wort: Liquid obtained from the mashed malt and malt adjuncts after these have been heated and their starches digested.

Basic Knowledge Questions

1. Describe a possible mechanism by which *Erwinia carotovora* causes soft rot of vegetables.
2. Name three types of fruit rot and the organisms responsible.
3. What are the concerns over the safety of fresh-cut produce packaged under vacuum? How can these concerns be addressed/minimized?
4. Compare the production of sour dough bread vs. idli bread, including the organisms used and the role that each plays in fermentation.
5. Describe the process of brewing beer and ale, including the enzymes involved.
6. What is "malo-lactic" fermentation, and why is it important to the wine industry?

7. Compare production of ogi, gari, bongkrek, and ontjom in terms of the steps required and organisms of importance.

Critical Thinking Questions

1. During the spoilage of fruit, explain the sequence of events that may take place in terms of the types of organisms that would be found, the order in which each type would affect the fruit, and explain your reasoning.
2. You are wanting to develop a new fermented beverage from black beans. List the steps you would follow and include the types of organisms you would use and the reasoning behind your decisions.

CHAPTER 9

Miscellaneous Food Products

Learning Objectives

The reader will be able to:

1. Describe the types of microorganisms found in delicatessen foods and how they cause spoilage of such products.
2. List and define the structural components of eggs, and describe the types of microorganisms that cause spoilage.
3. Describe how mayonnaise is made and discuss the types of microorganisms that cause spoilage in this product.
4. Describe the types of organisms and the spoilage that is seen in bakery products, frozen meat pies, candy, spices, and nuts.
5. Describe what types of dehydrated foods are available, the types of microorganisms that cause spoilage, and how they do so.
6. Define foods used for medical purposes and discuss the types of spoilage that can be observed, including the types of microorganisms responsible.
7. Define single-cell protein, provide a justification for its need, and discuss the types of organisms and products that are used, and the safety and nutritional quality of these.

Chapter Outline

1. Delicatessen and related foods.
   a) Salads and sandwiches.
   b) Cream pies and bread/biscuit/pie dough
2. Eggs.
   a) Structure of eggs
   b) Egg contamination and types of microorganisms present
3. Mayonnaise and salad dressing.
   a) Formulation
   b) Types of organisms present
4. Cereals, flour, and dough products.
5. Bakery products.
6. Frozen meat pies.
7. Sugars, candies, and spices.
8. Nutmeats.
9. Dehydrated foods.
10. Medical foods.
11. Single-cell protein.
    a) Definition
    b) Rationale
    c) Organisms
    d) Products
    e) Nutrition and safety

Terms and Definitions

Albumen: Multi-layer component of eggs, consisting mainly of water and proteins that surround the yolk.

Amylase: Enzyme produced by some microorganisms which breaks down starch into individual glucose molecules. These can be easily metabolized by the organism for growth and cell division.

Avidin: A protein found in the albumen of eggs, it binds the vitamin biotin.

Biotin: Water-soluble vitamin, found in eggs and many other foods. It acts as a coenzyme in carboxylation (carboxyl-group-adding) reactions in cells.

Black Rot: Spoilage condition of eggs caused by *Proteus*, *Aeromonas*, *Pseudomonas* bacteria.

Chalazae: A pair of spiral bands of albumen-like substance in the white of eggs. They extend from the ball of the yolk and are attached at each end of the egg to the vitelline membrane.

Colorless Rot: Spoilage condition of eggs caused by *Pseudomonas* and *Acinetobacter* bacteria.

Conalbumin: Synonymous with ovotransferrin, is one of the proteins found in the albumen of eggs. It consists of a peptide chain linked to iron.

Custard Rot: Spoilage condition of eggs caused by *Proteus vulgarius* and *P. intermedium*.

Emulsion: An aqueous system that contains two liquids that are not soluble in each other, one of which is dispersed in the other as droplets.

Lipolytic: Quality of some microorganisms that enable them to break down lipids through enzymatic action.

Lysozyme: Enzyme found in egg white and human tears, destructive to gram-positive bacteria.

Mucin: Glycoproteins found in secretions of many organisms, as well as tissues of man and animals.

Mustiness: Spoilage condition of eggs caused by *Pseudomonas graveolens* and other bacteria.

Ovoflavoprotein: Molecule that binds the vitamin riboflavin in eggs.

Ovotransferrin: Protein molecule that chelates metal ions such as iron, located in the albumen of eggs, synonymous with conalbumin.

Pink Rot: Spoilage condition of eggs caused by *Pseudomonas* bacteria.

Pinspots: Spoilage of eggs caused by molds, typified by appearance of mycelia in the inside of the egg, visible by candling.

Proteolytic: Quality of some microorganisms that enable them to break down proteins through enzymatic action.

Red Rot: Spoilage condition of eggs caused by some *Serratia* species.

Ropiness: Spoilage condition of homemade bread caused by the bacterium *Bacillus mesentericus*.

Rotting: Spoilage of eggs caused by bacteria, i.e. "green rot," "pink rot," "colorless rot," and "custard rot."

Shiga-like Toxin: Toxin that is produced by certain bacteria, typically pathogenic strains of *Escherichia coli* such as O157:H7. Toxin is similar in structure to Shiga toxin produced by the bacterium *Shigella dysenteriae*.

Vitelline Membrane: A membrane enclosing the ball of the yolk of eggs.

Basic Knowledge Questions

1. Draw a picture of the cross-section of an egg, labeling each structure/layer appropriately.
2. How do microorganisms enter eggs and what factors affect this?
3. Describe how mayonnaise spoils.
4. What properties of bakery products make them more prone to mold growth than other foods?
5. How does *Leuconostoc* spoil candy products?
6. At what stages during processing are contaminants introduced into nuts?

7. List the advantages of the use of single-cell protein as a source of protein for human consumption.

Critical Thinking Questions

1. Knowing what you now know about some of the components of eggs, discuss how you could use these to improve the microbial quality of salads.
2. After spending time in the poorest third-world country in the world, you decide to use your knowledge of single-cell protein to design a system of food production that will ease the malnourishment of that country's citizens. Describe the factors you need to consider and how you could introduce such products into the diet.

CHAPTER 10

Culture, Microscopic, and Sampling Methods

Learning Objectives

The reader will be able to:

1. Describe the standard plate count method, including new technologies for the processing of multiple samples.
2. Discuss membrane filtration techniques.
3. Describe microscope colony count methods.
4. Describe the agar droplet method and its applications.
5. Describe and compare three major dry-film-based methods.
6. Briefly define the Most Probable Numbers method and how it is applied for enumeration of microorganisms in foods.
7. Describe dye reduction and roll tube methods.
8. Describe direct microscopic count methods, including a discussion on distinguishing viable from nonviable cells.
9. List and describe several methods for the evaluation of surface contamination.
10. Define microbial injury and stress, and discuss how these processes occur, how the cell may repair the damage and recover, and how some cells can remain viable but not culturable.

Chapter Outline

1. General overview.
2. Conventional standard plate count.
   a) Factors to consider.
   b) Homogenization of food samples.
   c) The spiral plater.
3. Membrane filters.
   a) Definitions.
   b) Direct epifluorescent filter technique.
   c) Microcolony-DEFT.
   d) Hydrophobic grid membrane filter.
4. Microscope colony counts.
5. Agar droplets.
6. Dry film and related methods.
   a) Petrifilm
   b) Redigel
   c) SimPlate
7. Most probable numbers.
8. Dye reduction.
9. Roll tubes.
10. Direct microscopic count.
    a) Conventional method.
    b) Slide method.
    c) Howard mold counts.
11. Microbiological examination of surfaces.
    a) Swab/swab-rinse methods.
    b) Contact plate.
    c) Agar syringe/"agar sausage" methods.
    d) Other surface methods.
       1) Direct surface.
       2) Sticky film.
       3) Swab/agar slant.

4) Ultrasonic devices.
5) Spray gun.

12. Metabolically injured organisms.
    a) Definition of injury and stress.
    b) Recover/repair.
    c) Mechanism.
13. Viable but nonculturable organisms.

Terms and Definitions

Acridine Orange: Fluorescent dye which penetrates microbial cells.

Autoclaved: Refers to the condition of a material after it has been subjected to heating using an autoclave device to achieve sterility. In this procedure, the material (i.e. food or medium) is placed inside a chamber and heated with steam under pressure for varying times at temperatures of 250°F or more.

Beta-Glucuronidase: Enzyme produced by certain microorganisms, such as most strains of *Escherichia coli*, that breaks down glucuronic acid.

Lag Phase: Phase of bacterial growth in which cells are actively undergoing metabolic reactions, chiefly synthesizing macromolecules for cell growth, but no cell division takes place. It is during this phase that injured cells undergo repair.

Methylene Blue: Dye used as an indicator of the degree of anaerobiosis of a medium, with the dye turning from blue to colorless upon reduction of the medium when anaerobic conditions have been achieved. Such conditions can be caused by chemical means or by depletion of oxygen due to microbial growth.

Monoclonal Antibody: Antibody molecules produced by a single B-cell of an animal or person. Monoclonal antibodies are identical to each other, and are specific for a unique antigenic determinant, such as a particular component of the cell membrane of a given bacterium. They are useful for detecting the presence of particular microorganisms in food samples.

Protein A: Protein which binds nonspecifically to antibodies on their Fc portion, serving as an anchor prior to exposing them to an antigen, such as a bacterial cell produced by *Staphylococcus aureus*.

Resazurin: Dye used as an indicator of the degree of anaerobiosis of a medium, with the dye turning from pink to colorless upon reduction of the medium when anaerobic conditions have been achieved. Such conditions can be caused by chemical means or by depletion of oxygen due to microbial growth.

Quebec Colony Counter: Device equipped with a lamp, designed to shine light through a petri plate on which microbial colonies are growing. It allows for low-level magnification of colonies for ease of counting.

Stomacher: A blender-like device in which food samples in buffer are homogenized by gentle pummeling rather than high-speed blending.

TCA Cycle: Also known as Krebs Cycle, it is the series of reactions that take place in most microbial cells during the metabolism of sugars, yielding approximately 36 molecules of ATP per molecule of glucose.

Triton-X: Detergent used to remove proteins and other organic materials from surfaces. Chemical name is polyoxyethylene p-t-octyl phenol.

Trypan Blue: An acid dye that stains mold and other colonies with a blue color.

Trypsin: Enzyme found in gastric juice that is used to break down proteins.

Basic Knowledge Questions

1. What factors should be considered in conducting standard plate counts of microbial contaminants?
2. What are the types of fluorescent microscopy methods that can be used to evaluate microorganisms? How do they work?
3. Compare and contrast the Redigel and SimPlate methods to enumerate bacteria. What are the advantages/disadvantages of each?
4. What method would you use to sample the surface of a food processing plant after it has been sanitized? Why?
5. What is microbial injury?
6. What factors affect the ability of microorganisms to repair after injury?
7. Define "culturable but not viable".

Critical Thinking Questions

1. It is known that the number of pathogenic organisms in most foods is very low, provided those products have not been temperature-abused. What method(s) would you consider using to enumerate such organisms? Why?
2. You have developed a new isolation medium for the hypothetical bacterium *Testina obligata*. Design an experiment that would test the ability of this new medium to successfully and accurately enumerate injured as well as healthy cells.

CHAPTER 11

Physical, Chemical, Molecular, and Immunological Methods

Learning Objectives

The reader will be able to:

1. Define impedance, microcalorimetry, and flow cytometry, and be able to cite examples of where and how these physical detection methods are used in the food industry.
2. Define and describe chemical methods for detection of microorganisms, comparing those based on enzymatic assays and those based on other technologies.
3. Describe serotyping and bacteriophage typing, and compare the two traditional assays for bacterial classification and identification.
4. Describe the basis for various DNA-based methods of fingerprinting microorganisms.
5. Describe and discuss immunological methods, making comparisons of application and limitations of each.
6. Explain the principle behind the ELISA method, and describe its application for the detection of various types of microorganisms and their toxins.

Chapter Outline

1. Physical Methods.
   a) Impedance and related methods
   b) Microcalorimetry
   c) Flow cytometry
2. Chemical Methods.
   a) Thermostable nuclease
   b) Limulus lysate for endotoxins
   c) Adenosine triphosphate measurement
   d) Radiometry
   e) Fluorogenic and chromogenic substrates
   f) *Lux* gene luminescence
   g) Ice nucleation assay
3. Methods for Characterizing and Fingerprinting Foodborne Organisms
   a) Serotyping
   b) Bacteriophage typing
   c) Nucleic acid (DNA) probes
   d) DNA amplification (Polymerase Chain Reaction)
   e) Multilocus enzyme electrophoresis typing
   f) Restriction enzyme analysis
   g) Random amplification of polymorphic DNA
   h) Pulsed field gel electrophoresis
   i) Restriction fragment length polymorphism
   j) Ribotyping
4. Immunological Methods
   a) Fluorescent antibody
   b) Enrichment serology
   c) Salmonella 1–2 test
   d) Radioimmunoassay
   e) ELISA
   f) Gel diffusion
   g) Immunomagnetic separation
   h) Hemagglutination

Terms and Definitions

ATP: Adenosine triphosphate, it is the primary source of energy for all living cells, microbial as well as others. It is produced by the metabolism of nutrients and is used to fuel catabolic and other reactions.

Antigen: Term used to refer to any molecule or particulate structure, considered foreign to a particular animal or human being. Antigens can be part of the cell wall of a bacterium, and when introduced into an animal, elicit the production of antibodies by the animal that are specific for and bind to that antigen.

Cation Exchange: Method used to separate proteins in a sample, in which those with a net positive charge will bind to a column containing a carboxyl group, whereas proteins with a negative charge will not. The positively-charged proteins are then released from the column by adding sodium chloride or another salt to the column. Sodium ions in the buffer will compete with the positively charged groups on the protein, and will bind to the column. Those proteins with a low-density of net positive charge will emerge first out of the column, followed by those with a higher charge density.

Coagulase: Enzyme produced by some microorganisms, such as *Staphylococcus aureus*, which is used as an indicator of the ability of the cells to produce enterotoxin.

Enthalpy: The amount of heat lost by a system to the environment.

Exothermic HPR: Rate at which microorganisms produce heat during growth, which can be correlated to the number of organisms present in the medium.

Fluorescein Isothiocyanate: Fluorescent stain which binds to protein.

Flow Cytometry: The measurement of individual cells as they flow through a liquid medium. Cells can be individually measured in terms of size, fluorescence, light-absorbing properties, and can be counted as they pass by a laser beam detector.

Flow Micro-Calorimetry: Microbial identfication method in which a flow-through vessel is used containing a seven-sugar medium. Microorganisms such as yeasts and lactic acid bacteria are allowed to grow, producing thermograms unique to each organism.

H Antigens: Components of the flagella of motile bacteria, such as salmonellae.

Hybridization: Method in which two single-strands of DNA are allowed to anneal, or bind, at complimentary base pairs along the strands. The degree of hybridization is used as an indicator of the similarity of the DNA strands. If these come from two separate organisms, hybridization is used as an indicator of the similarity of the two organisms.

IDT: Impedance detection time, refers to the time required to detect resistance to the flow of electricity in a medium. IDT can be correlated to the number of microorganisms present in a broth medium, with the shorter the IDT the higher the number of cells present.

Impedance: The resistance to flow of electrical current. It is the opposite of conductivity.

K Antigens: Components of the capsule of encapsulated bacteria, such as streptococci.

Lipopolysaccharide (LPS): Structural component of the bacterial gram-negative cell envelope. It is pyrogenic, able to induce fever in humans who have been infected. One of its components is Lipid A. Also known as endotoxin.

Lysate Protein: Protein found in the amoebocytes, or blood cells, of the horseshoe crab. It is used to detect the presence of endotoxin by the formation of a clot when the protein comes in contact with the LPS of gram-negative cells.

O Antigens: The O polysaccharide side chain that is part of the LPS of gram-negative bacteria. These are found on the outermost surface of the cell, typically protruding out of the outer cell membrane. Also known as "somatic antigens".

Phage: A virus that is specific and generally lethal to bacteria, which can be manipulated to act as a carrier and insert certain desirable genes into the cells.

Propidium Iodide: Fluorescent stain that binds to DNA.

Restriction Enzymes: Enzymes that will bind to specific sites along DNA molecules (i.e. specific base sequences), lysing or cutting the molecule at those sites. Also known as "restriction endonucleases".

Serotyping: The use of antibodies specific for particular organisms for the purpose of identification and classification according to differences on the molecules residing on the surface of cells (i.e. in cell membrane).

Thermostable Nuclease: Enzyme produced by *Staphylococcus aureus*, especially those cells that produce enterotoxin, as well as other organisms. It degrades DNA and is resistant to boiling for 15 minutes.

Vero Toxin: Toxin produced by some pathogenic strains of *Escherichia coli*, such as O157:H7. It causes hemorrhagic colitis in humans. Toxin was first identified as being cytotoxic to Vero tissue culture cells (African Green Monkey kidney cells), hence the terms "vero toxin" and "verocytotoxin."

Basic Knowledge Questions

1. Compare the use of thermostable nuclease and coagulase as indicators of the presence of enterotoxin-producing *S. aureus*.
2. Discuss the LAL and ATP assays for detection of microorganisms. What do these have in common? What are the limitations of each?
3. Name several of the chromogenic substrates used in culture media for detection of microorganisms. What is the difference between fluoro- and chromogenic substrates?
4. Describe the various types of antigens found on gram-negative bacteria.
5. List some applications of the polymerase chain reaction assay for detection of microorganisms in food.
6. Compare and contrast PFGE and RFLP. Which do you think gives a more accurate fingerprint of a particular microbial isolate? Why?

7. Describe the conventional and "sandwich" ELISA assays.

Critical Thinking Questions

1. You are given antibodies specific for *Salmonella* species, developed in goats. The only other reagent you have is antibodies developed in humans, specific for goat tissue, to which an enzyme is bound. Your job is to design an ELISA assay that will help you decide whether a certain food sample contains this organism. Draw a diagram depicting the steps you would need to follow in carrying out the assay.
2. Name the factors you would recommend that an industry person considers when making a decision regarding the type of test to use for the following purposes:
   a) Verifying that a sanitation procedure on a food contact surface has indeed been successful.
   b) Determining whether a bacterial isolate found on a certain food product is the same as the strains found in the food processing plant environment.

CHAPTER 12

Bioassay and Related Methods

Learning Objectives

The reader will be able to:

1. Describe the various types of whole animal assays, stating the application of each.
2. Describe bioassays that require surgical intervention of animals, such as the ligated ileal loop test and the RITARD test.
3. Describe the various types of tissue culture cells that can be used to detect specific pathogens and/or their toxins.

Chapter Outline

1. Whole-Animal Assays.
   a) Mouse lethality
   b) *Suncus murinus*
   c) Suckling (infant) mouse
   d) Rabbit and mouse diarrhea
   e) Monkey feeding
   f) Kitten (cat) test
   g) Rabbit and guinea pig skin tests
   h) Sereny and Anton tests
2. Animal Models Requiring Surgical Procedures.
   a) Ligated loop techniques
   b) The RITARD model
3. Cell Culture Systems.
   a) Human mucosal cells
   b) Human fetal intestine
   c) Human ileal and intestinal cells
   d) Guinea pig intestinal cells
   e) HeLa cells
   f) Chinese hamster ovary cells
   g) Vero cells
   h) Y-1 adrenal cell assay
   i) Other assays

Terms and Definitions

Buccal Mucosa:	The lining of the inside of the mouth.
Cereulide:	Toxin produced by some strains of the bacterium *Bacillus cereus*, causes vomiting.
Depilation:	Removal of hair, usually by shaving, to expose the surface of the skin prior to intradermal injection.
Erythema:	Reddening of the skin, usually as a result of injection of certain substances intradermally.
Emesis:	Vomiting.
Hepatocyte:	Liver tissue cells.
Intracardial (i.c.):	One method of injection, literally meaning "through the heart".
Intradermal (i.d.):	One method of injection, literally meaning "through the skin".

Intraluminal (i.l.): Method of injection, literally meaning "through the lumen," or the inside of the intestines.

Intraperitoneal (i.p.): One method of injection, literally meaning "through the wall of the peritoneum," the sac-like material that surrounds the abdominal cavity of the body.

Kanagawa Test: Used to characterize strains of *V. parahaemolyticus*. Lysis of human or rabbit red blood cells but not horse on Wagatsuma's agar denotes a K^+ strain..

Necrosis: Tissue death, caused by the lack of blood circulation.

Orogastric: Through the mouth and into the stomach.

Phagolysosome: Sac-like structure in the cytoplasm of phagocytic immune system cells such as macrophages. It is formed by the merging of two cytoplasmic structures: a phagosome (containing an invading microorganism that has been engulfed by the macrophage) and a lysosome (containing lysozyme and other lytic enzymes) for the purpose of digestion of the invading organism.

PEU: Plating efficiency unit, or the amount of enterotoxin that results in a 25% inhibition of 200 tissue culture cells inoculated with 100 uL of medium.

Scintillation Counter: Instrument that measures the radioactivity of a sample by measuring alpha and beta emissions.

Trypsin: Proteolytic enzyme found in the gastrointestinal tract that is used to activate the toxin of some types of *Clostridum botulinum*.

Basic Knowledge Questions

1. Compare the Anton and Sereny tests.
2. Why is the *Suncus murinus* test used?
3. Define the GW/BW ratio as used in the Suckling Infant Mouse assay.
4. Name and describe the various types of skin tests used for detection of toxin.
5. Describe the ligated ileal loop test.
6. Construct a table that lists at least three types of tissue culture cells and the pathogen or agent that each test detects.
7. What are Vero cells and what are they used for?

Critical Thinking Question

1. There is a new microorganism that has been linked to an outbreak of foodborne illness. Describe the types of tests that you would employ to determine how it causes disease.

CHAPTER 13

Food Preservation with Chemicals

Learning Objectives

The reader will be able to:

1. Describe the mode of action of benzoic acid, parabens, sorbic acid, propionates, sulfur dioxide, and sulfites.
2. Explain the uses of nitrites and nitrates, and describe the types of organisms that are affected by these compounds.
3. Define the "Perigo Factor".
4. Describe how nitrites can be used in combination with other antimicrobial agents.
5. Describe the antimicrobial effect provided by sugars and salts.
6. Describe how several ingredients, added to food products for other purposes, act to inhibit microorganisms.
7. Describe the mechanism of action of lactic and acetic acid, and list some of the applications of these antimicrobial compounds.
8. Define the term "bacteriocin," and list and describe several types of these compounds, and the organisms that produce them.
9. Describe the use of antifungal agents and fumigants for decontamination of plant products.

Chapter Outline

1. Benzoic Acid and the Parabens.
2. Sorbic Acid.
3. The Propionates.
4. Sulfur Dioxide and Sulfites.
5. Nitrites and Nitrates.
   a) Compounds used
   b) Organisms affected
   c) The Perigo factor
   d) Interaction with cure ingredients and other factors
   e) Nitrosamines
   f) Nitrite-sorbate and other nitrite combinations
   g) Mode of action
   h) Summary of nitrite effects
6. NaCl and Sugars.
7. Indirect Antimicrobials.
   a) Antioxidants
   b) Flavoring agents
   c) Spices and essential oils
   d) Medium-chain fatty acids and esters
8. Acetic and Lactic Acids.
9. Antibiotics and Bacteriocins.
   a) General considerations
   b) Nisin and other bacteriocins
   c) Monensin
   d) Natamycin
   e) Tetracyclines
   f) Subtilin
   g) Tylosin
10. Antifungal Agents for Fruits.
11. Ethylene and Propylene Oxides.

12. Miscellaneous Chemical Preservatives.
   a) Sodium diacetate
   b) Ethanol
   c) Diethylpyrocarbonate
   d) Dimethyl dicarbonate
   e) Acid anionic sanitizers
   f) Sucrose fatty acid esters
   g) Glucose oxidase

Terms and Definitions

Bacteriostatic: An agent or chemical that inhibits growth of bacteria without killing the cell.

Chelate: A compound that sequesters particular ions in solutions, binding strongly enough to prevent their interaction with other molecules. One example is EDTA, which chelates iron and other multivalent ions.

Endospore: Physical structure developed by spore-forming bacteria, in which several layers are added to the cell for protection in harsh environments (i.e. dry conditions, lack of nutrients). In this stage, the cell is in a "dormant" state with no growth taking place. The spore can germinate upon exposure to proper conditions, at which time the cell that results can resume active growth.

Fistulation: A surgical procedure in which a in-out port is implanted on the skin of an animal in order to access the contents of particular abdominal chambers, such as the stomach or rumen. With this implant, samples of fluid can be drawn from that chamber after specific feeding regimens are applied to the animal, for example.

Fungistat: A compound that inhibits growth of fungal cells without causing death.

Germination: Phase in the life cycle of spore-forming organisms. It involves the shift from the spore-form to the vegetative cell form of the organism. Germination is often triggered in the spore by exposure to a short-term stress, such as mild heating, or by exposure to environmental conditions suitable for growth, such as the renewed availability of nutrients.

GRAS: Status given to some food ingredients, classifying them as generally recognized as safe for human consumption. Thus, testing and approval by the Food and Drug Administration is not required for these to be used. Examples are sugar and salt.

Ionophores: Small hydrophobic molecules that dissolve within lipid bilayers of membranes, acting as channels to allow the diffusion of ions into or out of the cell. Some antibiotics, such as valinomycin, act to transport potassium ions, disrupting transport mechanisms of the bacterial cell.

pKa: The pH of a weak acid in which half of the molecules are in the undissociated state and the other half are in the dissociated state.

Phosphoroclastic: Refers to a chemical reaction in which phosphate is removed from one molecule and used to phosphorylate another, such as in the conversion of ATP from ADP.

Plasmolysis: The lysis of microbial cells as a result of osmotic imbalance that causes the cells to lose water to the point of shrinkage (dehydration). It is the result of having a higher concentration of solutes in the medium compared to the inside of the cell.

PMF: Proton Motive Force, refers to the force that drives the movement of compounds in and out of the cell. It is dependent on the establishment of a charge differential, or gradient, across the cell membrane, in which positively charged ions (protons) are on the outside of the cell.

Undissociated: Status of a molecule in which hydrogen ions are no longer bound to the rest of the molecule. The molecule is then said to be in the oxidized, charged state, containing one unpaired electron. Weak acids are undissociated when the pH of the medium is below the pKa of the acid.

Viristatic: A compound that inhibits the proliferation of viruses without necessarily causing death.

Basic Knowledge Questions

1. Describe how spore-forming bacteria germinate, and how certain antimicrobial agents can inhibit this process.
2. Describe the mode of action of sulfites to inhibit microorganisms.
3. What is DNFH and how is it formed?
4. Define the "Perigo Factor", and give an example of it.
5. What is the mode of action of nitrite in inhibiting germination of spores?
6. Name some spices and the antimicrobial compounds found in each.
7. What are the classes of bacteriocins?

Critical Thinking Questions

1. Do you think the "Perigo Factor" is a real phenomenon? How would you test its existence?
2. Why do you think bacteriocins are usually inhibitory to organisms that are closely related to the one producing them?

CHAPTER 14

Food Preservation with Modified Atmospheres

Learning Objectives

The reader will be able to:

1. Define the various types of modified atmosphere packaging systems.
2. Describe the primary effects of MAP on inhibition of microorganisms.
3. List and describe various food products that can be packaged with MAP.
4. Discuss the food safety considerations of MAP-packaged foods.
5. Describe the type of spoilage that results in MAP-packaged foods.

Chapter Outline

1. Definitions.
   a) Hypobaric (low pressure) storage
   b) Vacuum packaging
   c) Modified atmosphere packaging
   d) Equilibrium-modified atmosphere
   e) Controlled-atmosphere packaging or storage
2. Primary Effects of CO_2 on Microorganisms.
   a) General findings
   b) Mode of action
3. Food Products.
   a) Fresh and processed meats
   b) Poultry
   c) Seafoods
4. The Safety of MAP Foods.
   a) *Clostridium botulinum*
   b) *Listeria monocytogenes*
   c) Other pathogens
5. Spoilage of MAP and Vacuum-Packaged Meats.
   a) Determining factors
   b) Volatile components of vacuum-packaged meats and poultry

Terms and Definitions

CAP:	Controlled-atmosphere packaging, refers to the packaging of foods in a modified atmosphere within an impermeable package so that the composition of the atmosphere inside the package does not change during storage.
Captech Process:	Packaging system in which hygienic conditions are followed in the packaging of the food with either gas-permeable or impermeable film, and storage at −1.5°C.
EMA:	Equilibrium-modified atmosphere, refers to the packaging of foods in gas-permeable film or material. The food is either

gas-flushed prior to sealing of the package, or not flushed at all. Gas exchange is thus allowed to take place between the food tissue and the outside environment.

Hyperbaric: Condition where the pressure is higher than atmospheric levels (i.e. 1 atm).

Hypobaric: Condition where the pressure is lower than atmospheric levels (i.e. 1 atm).

OTR: Oxygen transmission rate of a packaging material, usually expressed as volume of oxygen transmitted through the package per a specific area of the surface of the package within a given period of time, at a given temperature of storage, and under a specific relative humidity. For example: 8 mL/m^2/24 h/37°C/70% R.H.

Periplasm: Fluid space located between the inner and outer membranes of gram-negative bacteria. It contains several types of molecules, primarily proteins.

Basic Knowledge Questions

1. List some of the historical events surrounding the development of modified atmosphere packaging technology.
2. What are the two types of modified atmosphere packaging?
3. List some of the primary effects of gas-flushing on microorganisms.
4. List the hurdles against *Clostridium botulinum* growth that are inherent in modified atmosphere packaged foods.
5. What are the concerns regarding *Listeria monocytogenes* and modified atmosphere-packaged foods?
6. List the determining factors that guide the types of microorganisms that may be found in modified atmosphere-packaged foods.
7. Describe the products of spoilage of modified atmosphere-packaged meats, including the types of organisms responsible.

Critical Thinking Question

1. What are the possible mechanisms at play in inhibition of microorganisms by CO_2? How would you test your hypothesis?

CHAPTER 15

Radiation Preservation of Foods and Nature of Microbial Radiation Resistance

Learning Objectives

The reader will be able to:

1. Define the term "radiation" and describe the units employed for measuring dose.
2. Describe the various types of radiation that could be used for decontamination of foods.
3. Describe the effect of various parameters inherent to microorganisms on ability of irradiation to inactivate them.
4. Describe the effect of various parameters not related to the microorganism (packaging atmosphere, temperature, food composition) on microbial inactivation by irradiation.
5. Describe the types of foods most suitable for irradiation.
6. Describe how foods should be processed prior to irradiation.
7. Describe the process of irradiation, as it applies to the use of specific sources (i.e. electrons, gamma rays, X rays).
8. Define the terms "radappertization," "radurization," and "radicidation."
9. Discuss the legal status of food irradiation in the U.S.
10. Discuss the effect of irradiation on the quality of foods, and how detrimental effects can be minimized and/or prevented.
11. Discuss the topic of microbial resistance to irradiation, citing specific microbial examples.

Chapter Outline

1. Definitions and Units of Measure.
2. Characteristics of Radiations of Interest in Food Preservation.
   a) Ultraviolet light
   b) Beta rays
   c) Gamma rays
   d) X rays
   e) Microwaves
3. Principles Underlying the Destruction of Microorganisms by Irradiation.
   a) Types of organisms
   b) Numbers of organisms
   c) Composition of suspending menstruum
   d) Presence or absence of oxygen
   e) Physical state of food
   f) Age of organisms
4. Processing of Foods for Irradiation.
   a) Selection of foods
   b) Cleaning of foods
   c) Packing
   d) Blanching or heat treatment
5. Application of Radiation.
   a) Gamma radiation
   b) Electron beam/accelerated electrons
6. Radappertization, Radicidation, and Radurization of Foods.
   a) Definitions
      1) Radappertization

2) Radicidation
3) Radurization
7. Legal Status of Food Irradiation.
8. Effect of Irradiation on Food Quality.
9. Storage Stability of Irradiated Foods.
10. Nature of Radiation Resistance of Microorganisms.
   a) Biology of extremely resistant species
   b) Apparent mechanisms of resistance

Terms and Definitions

Beta Radiation: Electrons emitted by a radioactive isotope during its decay, such as cobalt-60.

Cold Sterilization: Processing treatment that causes no noticeable increase in temperature of the product. Term is often used to refer to ionizing radiation.

Electromagnetic Spectrum: Spectrum of all the electromagnetic energy present in nature, ranging from communication bands to gamma rays, and including visible light, microwaves, and ultraviolet radiation.

Electron Volt: Unit of measure of energy, described as the energy gained by an electron as it travels through a 1 volt potential.

Gamma Rays: Photons of electromagnetic energy emitted by a radioactive isotope during its decay, such as cobalt-60.

Gray: Unit of measure for absorbed dose. It is equivalent to 100 rads, and to 1 joule of energy/kg of matter.

Ionizing Radiation: Type of radiation in which matter is exposed to high levels of energy within the electromagnetic spectrum, causing ionization of some of its molecules.

Microwaves: Type of energy produced when electrically neutral foods are placed in an electromagnetic field of a particular wavelength. It causes the molecules in food to oscillate, creating friction and thereby heat.

Rad: Unit of measure for absorbed dose. It is equivalent to 0.01 joule of energy/kg of matter.

Radappertization: Application of ionizing radiation to achieve commercial sterility, with typical doses ranging from 30 to 50 kGy.

Radiation: Emission of energy and its propagation through space or matter (i.e. food).

Radicidation: Application of ionizing radiation to achieve pasteurization, with typical doses ranging from 2 to 10 kGy.

Radiolysis: The lysing effect of chemical bonds as caused by exposure of a material to radiation.

Radiolytic Products: Products of the lysis of chemical bonds by application of various food processing treatments such as ionizing radiation, heat, and freezing. For example, oxygen radicals derived from the lysis of water molecules.

Radurization: Application of ionizing radiation to achieve low-grade pasteurization, with typical doses ranging from 0.75 to 2.5 kGy.

X Rays: Photons of energy emitted as a result of collision of accelerated electrons with a heavy metal.

Basic Knowledge Questions

1. What are the units used to express dose and energy?
2. List the types of radiation found in the electromagnetic spectrum of energy.

3. How are microorganisms affected by ionizing radiation?
4. How do the following parameters affect microbial inactivation in foods by ionizing radiation?
   a) Oxygen content of the package
   b) Physical state of the food
   c) Age of the microorganisms
5. List the advantages of irradiating foods by accelerated electrons vs. gamma rays.
6. What is the legal status of red meat irradiation in the U.S.?
7. What procedures can be used to reduce or prevent any damage to food quality by irradiation?

Critical Thinking Questions

1. You need to irradiate a large quantity of a product that is less than one inch thick. What irradiation source would you use, and why?
2. Under what circumstances or conditions would you expect a foodborne pathogen to develop resistance to irradiation?

CHAPTER 16

Low-Temperature Food Preservation and Characteristics of Psychrotrophic Microorganisms

Learning Objectives

The reader will be able to:

1. Define the various classifications for organisms that can grow or tolerate cold temperatures.
2. Describe the minimum temperature requirements for psychrotrophs and other types of organisms.
3. Describe how foods are frozen.
4. Understand the difference between slow and quick freezing, and how these affect food quality and storage stability.
5. Describe the effect of freezing and thawing on survival of microorganisms.
6. Discuss the various characteristics of psychrotrophic and psychrophilic microorganisms in terms of their physiology and ability to conduct specific functions.
7. Discuss the effect of low temperature on microbial physiology and growth.
8. Explain why psychrotrophic organisms have a low tolerance for temperatures above their optimal range for growth.

Chapter Outline

1. General Considerations.
2. Definitions.
3. Temperature Growth Minima.
4. Preparation of Foods for Freezing.
5. Freezing of Foods and Freezing Effects.
   a) Quick freezing
   b) Slow freezing
6. Storage Stability of Frozen Foods.
7. Effect of Freezing on Microorganisms.
   a) Freezing
   b) Thawing
8. Some Characteristics of Psychrotrophs and Psychrophiles.
   a) Increase in unsaturated fatty acid residues
   b) Synthesis of high levels of polysaccharides
   c) Pigment production
   d) Differential substrate utilization
9. The Effect of Low Temperatures on Microbial Physiologic Mechanisms.
   a) Slower metabolic rate
   b) More efficient transport of solutes
   c) Larger cells
   d) More efficient synthesis of flagella
   e) Aeration effect
   f) Increased requirement for organic nutrients
10. Nature of the Low Heat Resistance of Psychrotrophs.

Terms and Definitions

Blanching: Process by which foods are exposed briefly to either boiling water or steam for the purpose of inactivating enzymes, among other reasons.

Bound Water: Water that is held by chemical bonds to other molecules, thus it is not available for microbial use.

Cryoprotectant: Chemical compound that lowers the freezing point of water.

Eurypsychrotroph: Psychrotrophic organism that does not form visible colonies until 6 to 10 days at 7°C.

Free Water: Water that is not bound to other molecules, thus it is available for microbial use.

Lipid Solidification Theory: Theory that seeks to explain the ability of some microorganisms to survive and grow at low temperatures. It states that at such temperatures, more unsaturated than saturated fatty acids are synthesized and incorporated into the cell membrane. The presence of these fatty acids make the membrane more fluid, preventing it from freezing.

Psychrophile: Microorganism that can grow from below zero to 20°C, with an optimum growth range of 10 to 15°C.

Psychrotroph: Microorganism that can grow between zero and 7°C, producing visible colonies within 7 to 10 days.

Stenopsychrotroph: Psychrotrophic organisms that form visible colonies within 5 days at 7°C.

Temperature Coefficient: Expressed symbolically as "Q_{10}", it is defined as the velocity of a chemical reaction at a given temperature + 10°C, divided by the velocity of the reaction at that same temperature.

Basic Knowledge Questions

1. List the purposes of blanching.
2. Compare and contrast quick vs. slow freezing in terms of their effect on food quality and on microorganisms.
3. Discuss the physiological effects that freezing has on microorganisms (not psychrotrophs or psychrophiles).
4. Describe the effect of freezing and thawing on food.
5. List at least three characteristics of psychrophilic and psychrotrophic microorganisms that enable them to grow at low temperatures.
6. How does fatty acid composition relate to the ability of cells to withstand low temperatures?
7. Explain the effect of low temperature on protein synthesis.

Critical Thinking Questions

1. Describe what happens during "cold shocking" of bacterial cells. Under what circumstances during food processing could such an event take place?
2. Explain how low temperatures affect the uptake of solutes by psychrotrophic organisms. What would happen to the uptake of solutes if such an organism were to suddenly be exposed to higher temperatures?

Chapter 17

High-Temperature Food Preservation and Characteristics of Thermophilic Microorganisms

Learning Objectives

The reader will be able to:

1. Define the terms "pasteurization" and "sterilization."
2. Discuss how various extrinsic and intrinsic factors can affect survival of microorganisms to heat treatments.
3. Discuss the phenomenon of heat resistance within the context of specific microorganisms and the mechanisms they employ to survive heating.
4. Define "D value," "z value," "F value," and other terms related to the measurement of heat resistance of microorganisms.
5. Define "thermophiles" and provide examples of such organisms.
6. Describe how thermophiles can survive high-temperature treatments, providing examples of the mechanisms and factors that affect this.
7. Discuss spoilage of canned foods, including foods that are of various acid content.

Chapter Outline

1. Types of High-Temperature Treatments.
   a) Pasteurization
   b) Sterilization
2. Factors Affecting Heat Resistance in Microorganisms.
   a) Water
   b) Fat
   c) Salts
   d) Carbohydrates
   e) pH
   f) Proteins and other substances
   g) Numbers of organisms
   h) Age of organisms
   i) Growth temperature
   j) Inhibitory compounds
   k) Time and temperature
   l) Effect of ultrasonics
3. Relative Heat Resistance of Microorganisms.
   a) Resistance of various types
   b) Spore resistance
4. Thermal Destruction of Microorganisms.
   a) Thermal Death Time
   b) D value
   c) z value
   d) F value
   e) Thermal Death Time curve
   f) 12D concept
5. Some Characteristics of Thermophiles.
   a) Growth
   b) Enzymes
   c) Ribosomes
   d) Flagella
6. Other Aspects of Thermophilic Microorganisms.
   a) Nutrient requirements
   b) Oxygen tension

c) Cellular lipids
d) Cellular membranes
e) Effect of temperature
f) Genetics

7. Canned Food Spoilage.
   a) Causes of spoilage
   b) Low acid foods
   c) Acid foods
   d) High acid foods

Terms and Definitions

Acid Foods: Foods that have a pH between 3.7 and 4.6.

"Bot Cook": Term for the heat treatment required to destroy 12 log of *Clostridium botulinum* spores in a food product.

Commercial Sterilization: Process by which foods are exposed to levels of heat that are high enough to destroy all vegetative cells, as well as spores of *Clostridium botulinum*, such that no microbial growth is observed during room-temperature storage.

D Value: The time of heating, usually expressed in minutes, required to decrease a microbial population by 90%, or by 1.0 log_{10} in a specific medium and at a specific temperature. It is an indirect measure of how heat resistant a microorganism is. It is also used to compare the heat resistance of two different organisms, with the higher the D value the higher the resistance. The D value can be calculated by taking the reciprocal of the absolute value of the slope of a "survivor curve." A survivor curve is generated by plotting the relationship between the number of cells in a medium vs. the time that the cells were heated at a given temperature.

F Value: The time, usually in minutes, at 250°F, required to destroy all spores or vegetative cells of a particular organism. It is obtained from a TDT curve by noting the specific temperature at which no survivors are observed. It is used as a way to measure the ease or difficulty of sterilizing a particular food product, with the higher the F value the more difficult it is to eliminate all cells of a specific organism in a specific food.

Flipper: Condition of a can in which one end becomes convex due to expansion of the metal as a result of excessive heat, physical damage, or pressure at the other end.

Hard Swell: Condition of a can in which both ends appear bulged, and neither can be dented by applying pressure. This is usually due to release of CO_2 and H_2S gas by spoilage microorganisms, or to formation of H_2 gas by a non-microbial chemical reaction between iron in the can and acidic compounds in the food.

High-Acid Foods: Foods that have a pH less than 3.7.

Leakage-Type: Type of spoilage of canned foods caused by nonspore-forming microorganisms. These organisms would not be able to survive the canning process, thus their presence is due to faulty seams and small holes in the can.

Low-Acid Foods: Foods that have a pH higher than 4.6.

Pasteurization: Process by which foods are exposed to levels of heat that are high enough to destroy all disease-producing vegetative cells, or to reduce the number of spoilage organisms.

Protoplast Dehydration: Reduction of the water content of the inside of a bacterial spore as a way of increasing the spore's resistance to heating. Diminution of water lowers heat conductivity thus protecting the spore from damage.

Soft Swell: Condition of a can in which both ends appear bulged, but which can be dented if pressed with moderate force.

Springer: Condition of a can in which one or both ends appear concave and spring out upon pushing, resulting in bulging.

Sterilization: Process by which foods are exposed to levels of heat that are high enough to destroy all viable microbial cells.

Thermal Death Time: Time, usually in minutes, necessary for a heat treatment to kill a specific number of cells at a specific temperature. Experiments are conducted to determine the D value at various temperatures for a particular organism, and the relationship between the $\log_{10}$ of each D value vs. temperature is plotted. From this curve one can obtain the "z value" of an organism.

Thermophile: Microorganism that grows at a minimum temperature of 45°C, with an optimum range of 50 to 60°C, and a maximum at 70°C or above.

UHT: Ultra-high temperature pasteurization, refers to a process whereby milk and milk products are exposed to very high temperatures (140 to 150°C) for a few seconds, rendering them commercially sterile.

Z Value: The temperature, in degrees Farenheit, necessary to decrease the TDT by 90%, or 1.0 $\log_{10}$. It is a constant that can be used, along with the D value of an organism at a particular temperature, to determine what the D value would be at a different temperature. It can be calculated by taking the reciprocal of the absolute value of the slope of a TDT curve. A TDT curve is generated by plotting the relationship between the $\log_{10}$ of D values obtained at various temperatures vs. the temperatures themselves.

Basic Knowledge Questions

1. How does the concentration of salt, fat, carbohydrates, and proteins affect microbial resistance to heat?
2. How does pH affect microbial resistance to heat?
3. How does the number and age of cells affect their survival to heat?
4. What physiological characteristics make some microbial cells more heat-resistant than others?
5. Define "D" and "z" values, both in mathematical and in practical terms.
6. What are the three types of enzymes found in thermophilic organisms?
7. Describe low -vs. high-acid foods, giving examples of organisms that can spoil each type as well as the condition of the product due to spoilage.

Critical Thinking Questions

1. Discuss the various theories explaining why bacterial spores are more resistant to heating than vegetative cells.
2. Assume that you obtained the following data after heating a bacterial spore suspension in cream-style corn:

Minutes	Number of survivors
0	10^8
20	10^6
40	10^4

a) Plot a survivor curve, properly label each axis, and calculate the D value.

b) Calculate the time you would have to heat the product in order to achieve a 12D process if you started with 10^5 spores.

CHAPTER 18

Preservation of Foods by Drying

Learning Objectives

The reader will be able to:

1. Define the various categories of dried foods, according to moisture content.
2. Describe how low-moisture foods are prepared, citing examples.
3. Describe how various types of organisms are affected by drying, and what causes microbial death.
4. Explain browning and how certain parameters can be manipulated to minimize this effect during drying.
5. Define intermediate-moisture foods (IMF) and describe how these are processed.
6. Discuss spoilage of IMF during storage, including microbial and nonmicrobial factors.
7. Define glass transition and explain how it can be used as an alternative to water activity measurements.

Chapter Outline

1. Categories of Foods.
2. Preparation and Drying of Low-Moisture Foods.
   a) Fruits and vegetables
   b) Meat
   c) Milk
   d) Eggs
   e) Freezing as a drying treatment
3. Effect of Drying on Microorganisms.
   a) Bacteria
   b) Yeasts and molds
   c) Death and injury
4. Storage Stability of Dried Foods.
   a) Browning
   b) Relative humidity
5. Intermediate-Moisture Foods.
   a) General definitions and history
   b) Preparation of IMF
   c) Microbial aspects of IMF
   d) Storage stability of IMF
   e) IMF and glass transition

Terms and Definitions

Alarm Water Content: The amount of water of a food, above which mold growth will take place.

Case-Hardening: A condition of vacuum-dried food in which a hard layer forms at the surface of the product, causing a slowing-down of the drying process, and of reconstitution at a later time.

Drum-Drying: A process in which milk-based

and other products, such as mashed potatoes, are dried by allowing it to coat the outside of stainless steel drums, which have been heated to 100–150°C with steam. Product dries into sheets, which peel off the surface of the drums and are cut into flakes as they drop into an auger at the bottom of the drum-dryer.

ERH: Equilibrium Relative Humidity, refers to the water that is present in a food product, which can be lost due to desorption. It can be defined mathematically as the ratio of the partial pressure of water vapor generated by a food sample, to the partial pressure of water vapor generated by pure water.

Freeze-Drying: A process in which egg-based products, coffee, juices, and other products that are deemed to be sensitive to heat are dried. Food is frozen inside a chamber. A vacuum is then applied, causing moisture in the food to sublime (i.e. goes directly from a solid to a gas). This results in drying of the product from the surface to the inside until the moisture content is below 5%. The food develops a sponge-like structure due to space-gaps caused by the escaping water molecules.

High-Temperature Vacuum-Drying: Method of drying particulate foods such as citrus juices and instant tea. The temperature of the food and the rate at which water is removed from it are controlled by heating it under a vacuum. The food is placed in a chamber on top of plates that are heated by circulating a heated fluid through them, causing moisture to escape. A vacuum is applied, helping to draw the moisture out more efficiently, exhausting it from the chamber to prevent it from condensing on the surface of the product.

Humectant: A compound that binds and retains water.

Hysteresis: The condition resulting from an excess in adsorbed moisture by a system above its equilibrium level.

Intermediate-Moisture Foods: Also known as IMF, refers to foods that have a moisture content between 15 and 50% and a water activity between 0.60 and 0.85. Examples are cakes, jams, fruit juice concentrates, and sweetened condensed milk.

Low-Moisture Foods: Also known as LM, refers to foods that have a moisture content of less than 25% and a water activity between 0 and 0.60. Examples are freeze-dried coffee and crackers.

Maillard Reaction: Also known as "non-enzymatic browning," it refers to the development of dark pigments as a result of reaction between aldehydes and amino groups of sugar and protein molecules. It is favored by high temperature and low moisture, since these increase the concentration of reactants. Maillard browning is seen to occur fastest when the moisture content of foods decreases to about 15–20% during drying, and is inhibited below 2% moisture.

Sorption Isotherm: The relationship between the amount of water either adsorbed or lost (desorbed) by a system (i.e. food), as a func-

	tion of its relative humidity or water activity. An example of a food undergoing adsorption: crackers inside a humid chamber. An example of a food undergoing desorption: fruit inside a dry chamber.
Sun Drying:	The process of drying foods by exposing them to sunlight for long periods of time. Raisins (dried grapes) are an example.
Water Dynamics:	Food components that respond to changes in the moisture and temperature of the food by changing their physical state, from a viscous glassy state to a rubbery structure. This transition occurs at a specific temperature, termed "T_g". The higher the moisture content, the lower the T_g, thus this value could be used as an alternative to water activity for the purpose of predicting microbial activity in foods.

Basic Knowledge Questions

1. Name some advantages of freeze-drying over high-temperature vacuum-drying.
2. Explain the relationship between water activity and spoilage of foods. Provide examples of microorganisms responsible for spoilage at various water activity levels.
3. List at least two strategies that can be used to minimize quality changes during drying.
4. Present the mathematical equation for calculating water activity, based on Raoult's law of mole fractions.
5. List the four techniques that can be used to change the water activity during production of intermediate-moisture foods.
6. Define Equilibrium Relative Humidity, and explain its relationship with the amount of dissolved solids in a food.
7. What is "glass transition"?

Critical Thinking Questions

1. Research has shown that microorganisms are more susceptible to the lowering of water activity in foods that have been prepared by adsorption, compared with foods that have been prepared by desorption. Provide a theory that would explain this phenomenon.
2. Compare freezing and drying of foods relative to the effect of these treatments on survival and growth of microorganisms, and on overall product quality.

CHAPTER 19

Other Food Preservation Methods

Learning Objectives

The reader will be able to:

1. Discuss alternative, or unconventional methods of food preservation.
2. Describe the principle behind high-hydrostatic pressure (HHP) processing of foods.
3. Explain how HHP affects food components.
4. Explain the mechanism of action of HHP against microorganisms, citing specific examples of how certain pathogens are affected.
5. Describe the process of applying pulsed electric fields in food processing.
6. Discuss the principle behind the antimicrobial effects of pulsed electric fields.
7. Discuss aseptic packaging as a food preservation technology, and how it is used in conjunction with conventional food processing treatments such as heat pasteurization.
8. Discuss manothermosonication, describing the principles behind the process and how it affects microbial cells.

Chapter Outline

1. High-Pressure Processing.
   a) Description of the process
   b) Some principles and effects of HHP on foods and organisms
   c) Effects of HHP on specific foodborne organisms
2. Pulsed Electric Fields.
   a) Mode of action
   b) Effects on microorganisms
3. Aseptic Packaging.
   a) Description of the process
   b) Applications
4. Manothermosonication (Thermoultrasonification).
   a) Definitions
   b) Effect on microbial contaminants in foods

Terms and Definitions

Aseptic Packaging:	Refers to the packaging of sterile products into sterile containers. Sterility is maintained by conducting all activities in a "clean room" environment in order to prevent introduction of microbial contaminants into the food.
Baroresistant:	Describes organisms that are relatively resistant to high pressures (above atmospheric levels).
Hydrostatic:	Refers to the uniform distribution of a treatment in a men-

struum. When applied as part of the high-hydrostatic pressurization of foods, the pressure increases instantaneously and is distributed evenly throughout the product.

Nisin: A bacteriocin produced by some strains of *Lactococcus lactis*, which is inhibitory to certain microorganisms.

PediocinAcH: A bacteriocin produced by the bacterium *Pediococcus acidilactici*, and which is inhibitory to certain organisms.

Plasmin: Enzyme found in milk which hydrolyzes casein proteins. It is inactivated by pasteurization.

Sonication: Treatment by high-frequency sound waves. When applied simultaneously with heat, the combined treatment is known as "manothermosonication".

Basic Knowledge Questions

1. Cite some examples of other processes that have been used in a hurdle approach in conjunction with HHP to reduce microbial contaminants in foods.
2. How does microbial resistance to HHP relate to resistance to other treatments?
3. Describe the effect of water activity on effectiveness of HHP in reducing microorganisms.
4. Describe some examples of how HHP can affect the quality of vegetables.
5. What causes cell death by PEF?
6. List 3 advantages of aseptic packaging over other processing methods.
7. What is manothermosonication?

Critical Thinking Question

1. Experiments have shown that bacteria differ in terms of their ability to survive HHP treatment. Refer to the list of organisms on page 378 and provide a theory that may help explain this phenomenon.

CHAPTER 20

Indicators of Food Microbial Quality and Safety

Learning Objectives

The reader will be able to:

1. Discuss the types of organisms and assays that can be used to indicate product quality, listing the criteria that should be used in selecting such organisms.
2. List the criteria used for selecting indicator organisms for food safety.
3. Define the term "coliform" and describe the general characteristics of this class of organisms (i.e. growth, distribution, etc).
4. Discuss the use of coliforms as indicator organisms of food safety, including limitations that should be considered.
5. Describe enterococci and discuss their use as indicator organisms.
6. Discuss the use of bifidobacteria and coliphages as indicators of food safety.
7. Discuss the correct use of fecal coliforms as indicators of food safety, including the potential for their overuse.
8. Define predictive microbiology and discuss its use as a method to predict spoilage and/or growth of pathogenic microorganisms in foods.

Chapter Outline

1. Indicators of Product Quality.
   a) Criteria for indicator organisms
   b) Examples of indicators
2. Indicators of Food Safety.
   a) Criteria for indicator organisms,
   b) Coliforms
      1) Strains
      2) Growth
      3) Detection and enumeration
      4) Distribution
      5) Coliform criteria and standards
      6) Some limitations for food safety use
   c) Enterococci
      1) Historical background
      2) Classification and growth requirements
      3) Distribution
      4) Relationship to sanitary quality of foods
   d) Bifidobacteria
   e) Coliphages
      1) Assay
      2) Utility for water
      3) Utility for foods
3. Possible Overuse of Fecal Indicator Organisms.
4. Predictive Microbiology/Microbial Modeling.

Terms and Definitions

Citrate Test: Assay used to indicate whether a particular organism can utilize citrate as a sole carbon source for growth. Most *Escherichia coli* strains are unable to do so, thus yield a negative reaction with this test.

Coliform: A gram-negative, nonspore-forming rod that can ferment lactose, producing gas within 48 hours at 35°C. Such organisms are used as indicators of the potential for contamination of food and water with feces, and thus of the presence of intestinal pathogens.

Coliphage: A virus that infects coliform bacteria. Coliphages are found in close proximity with coliforms, thus have been used as indicators of the presence of these bacteria.

Diacetyl: A compound produced from fermentation of milk by heterofermentative lactic acid bacteria. It imparts specific flavor notes and buttermilk aroma to fermented products. It can also be used as an indicator of spoilage of foods where fermentation by bacteria is considered undesirable (i.e. fruit juices).

F^+ Strains: Cells that have the F factor, piece of DNA containing genes that enable the cell to transfer genetic information to other cells through conjugation. This is accomplished by the synthesis of a structural appendage known as a pilus, which serves as a conduit to physically exchange genetic material (in the form of plasmids) with cells of the same or similar species. F^+ strains are also referred to as "male" cells. If the F factor becomes integrated into the chromosome of the F^+ cell, the cell becomes an Hfr strain, enabling it to begin preparations for the conjugation event.

Fecal Coliform: A gram-negative, nonspore-forming rod that can ferment lactose, producing gas within 48 hours at 44.5°C. Such organisms are used as indicators of the potential for contamination of food and water with feces, and thus of the possible presence of intestinal pathogens. It is considered by some as a more accurate indicator of fecal contamination than coliforms.

HACCP: Hazard analysis critical control point, refers to a management system that seeks to control foodborne hazards through prevention. It can be applied during production, processing, distribution, and preparation of food, with the goal of minimizing or eliminating health hazards to the consumer.

Hfr Strain: A cell that has integrated the F factor into its chromosomal DNA. Such a strain is able to initiate conjugation, and thus transfer other plasmids it may have to other cells.

Indole Test: Assay used to indicate the production of indole by a microorganism during metabolism. Typically, *Escherichia coli* will provide positive results while other coliforms such as *Enterobacter* will not.

Methyl Red Test: Assay used to indicate whether acidic compounds have been produced by a microorganism, resulting in the lowering of pH to below 4.5. Most strains of *Escherichia coli* are positive for this reaction.

Murein: Molecule that forms part of the bacterial cell wall.

PFU: Plaque-forming unit, it refers to the empty space left on the surface of an agar plate by a colony of bacteria after being destroyed by phages. The presence of PFUs is an indication that the bacteria on the plate were destroyed by phages, thus it serves to identify those bacteria to group or species.

Scombroid Poisoning: A foodborne illness caused by consumption of seafood containing high levels of histamine. This compound is produced by microorganisms during metabolism of the amino acid histidine.

Selective Culturing: Culturing method that utilizes a medium that contains certain ingredients that will inhibit specific organisms, allowing (selecting for) the growth of others. An example is Brilliant Green Lactose Broth, which contains the dye brilliant green that inhibits most gram-positive bacteria, thus selecting for gram-negative organisms, mostly of intestinal origin.

Voges-Proskauer Test: Assay used to indicate whether a certain microorganism produces acetic methylcarbinol during metabolism. *Escherichia coli* strains are unable to produce this compound.

Basic Knowledge Questions

1. List all the criteria that should be used for selecting a good indicator organism for food quality, and one for food safety.
2. Explain how spoilage of food can be determined without microbial testing.
3. List the four genera of coliform bacteria.
4. What are the two basic questions that should be considered in establishing enumeration standards for *E. coli* and/or coliforms in food?
5. List the main characteristics of enterococci that make them useful indicator organisms.
6. What are coliphages most useful for?
7. Discuss the relationship between F^+ bacterial strains and coliphages.

Critical Thinking Questions

1. Is it a good idea to set limits on the number of coliforms that can be found in meat? Why or why not?
2. What indicator organisms would you choose to determine whether fecal contamination of freshwater catfish had occurred? Defend your choices.

CHAPTER 21

The HACCP System and Food Safety

Learning Objectives

The reader will be able to:

1. Discuss the concept of the Hazard Analysis Critical Control Point system (HACCP).
2. Describe the prerequisite programs that must be in place before HACCP can be implemented.
3. Define various terms related to the HACCP system, such as "hazard", "CCP", etc.
4. List the seven principles of HACCP.
5. Describe the seven principles of HACCP.
6. Discuss how the seven principles would be applied in the manufacture of a particular example, cooked beef patties.
7. Discuss some limitations of the HACCP system.
8. Define "microbiological criteria" for foods.
9. Discuss the various classes of sampling plans.
10. List some of the criteria that have been established for various food products.

Chapter Outline

1. Hazard Analysis and Critical Control Point System.
   a) Introduction and history
   b) Prerequisite programs
   c) Definitions
   d) HACCP principles
      1) Principle 1: assess hazards and risks
      2) Principle 2: determine CCPs
      3) Principle 3: establish critical limits
      4) Principle 4: establish procedures to monitor CCPs
      5) Principle 5: establish corrective actions
      6) Principle 6: establish procedures for verification
      7) Principle 7: establish effective recordkeeping systems
   e) Flow diagrams
   f) Application of HACCP principles
      1) Principle 1: hazards and risks
      2) Principle 2: CCPs
      3) Principle 3: critical limits
      4) Principle 4: monitoring CCPs
      5) Principle 5: corrective actions
      6) Principle 6: verification
      7) Principle 7: recordkeeping
   g) Some limitations of HACCP
2. Microbiological Criteria.
   a) Definitions
   b) Sampling plans
   c) Microbiological criteria and food safety
   d) Microbiological criteria for various products
   e) Other criteria/guidelines

Terms and Definitions

Advisory Criterion: A microbiological specification for an end product which is intended to increase assurance that hygienic significance has been achieved. It may also refer to a microbiological guideline applied in a food establishment at a point during or after processing in order to monitor hygiene.

Control Point: Any point in a specific food system where loss of control does not lead to an unacceptable health risk.

Critical Control Point: CCP, refers to any point or procedure in a food system where control can be exercised and a hazard can be minimized or prevented.

Critical Limit: One or more prescribed tolerances that must be met to ensure that a CCP effectively controls a microbiological health hazard.

CCP Decision Tree: A sequence of questions to assist in determining whether a control point is a CCP.

Corrective Action: Procedures that must be followed when a deviation of the critical limit occurs.

HACCP Plan: A written document that delineates all the procedures to be followed in implementing the HACCP system.

Hazard: Any biological, chemical, or physical property of a food that may cause an unacceptable risk to the health of the consumer. Examples are: bacteria (biological), pesticides (chemical), glass (physical).

Mandatory Criterion: A microbiological standard that normally contains limits for pathogenic organisms of public health significance only. However, limits could also be set for nonpathogenic organisms.

Monitoring: A planned sequence of observations or measurements designed to produce an accurate record and to determine if a critical limit is being violated.

Standard: Part of a law or regulation that is enforceable by a particular regulatory agency. In the case of microbial standards, it is the limit of the number of organisms, or of positive samples, allowable by that agency.

Validation: The collection and evaluation of scientific and technical information to determine whether the HACCP plan, when properly implemented, will effectively control the hazards.

Verification: Methods, procedures, and tests used to determine whether the HACCP plan is being properly implemented.

Basic Knowledge Questions

1. What are the three categories of hazards one could find in food?
2. List the seven principles of HACCP.
3. What are some examples of verification documents?
4. List some of the limitations of HACCP.
5. List the 5 components of the definition of a microbiological criterion, according to the Codex Alimentarius Commission.
6. What are the main differences between a 2-class and a 3-class sampling plan?
7. List the types of organisms for which standards have been established.

Critical Thinking Questions

1. Why is recording the time/temperature of processing of a food product a better monitoring method than enumerating the organisms that may have survived the process?

2. Which of the following sampling plans is more conservative in terms of not allowing hazardous product from reaching the consumer? Why?
   Plan A: n=10, c=5 and $m=10^3$
   Plan B: n=1000, c=100 and $m=10^4$

CHAPTER 22

Introduction to Foodborne Pathogens

Learning Objectives

The reader will be able to:

1. Explain how foodborne pathogens are transmitted via the "fecal-oral" route.
2. List and describe the various hurdles that foodborne pathogens must overcome to cause disease in humans.
3. Discuss the acid tolerance response.
4. Discuss how gram-positive bacteria cause disease, listing examples of significant foodborne pathogens.
5. Describe the general mechanisms by which *Listeria monocytogenes* causes illness.
6. Discuss the virulence factors of salmonellae, *Escherichia coli*, yersiniae, shigellae, and vibrios, explaining how these serve to cause disease.

Chapter Outline

1. Introduction.
   a) The fecal-oral transmission of foodborne pathogens.
   b) "Universal" requirements
   c) Attachment sites
   d) Sigma factors and the acid tolerance response
2. Pathogenesis.
   a) Gram-positive bacteria
      1) Examples
      2) *Listeria monocytogenes*
   b) Gram-negative bacteria
      1) *Salmonella*
      2) *Escherichia coli*
      3) Yersiniae
      4) Shigellae
      5) Vibrios
3. Summary.

Terms and Definitions

(note that the following do not include terms compiled in the textbook in Table 22–3)

Adenylate Cyclase: Enzyme that catalyzes a reaction in which ATP is broken down to cAMP. Certain bacterial toxins, such as cholera toxin, activate this reaction, resulting in accumulation of cAMP by the intestinal tissue cell. This disrupts the transport of ions across the cell membrane, causing efflux of water to the outside of the cell. The resulting effect is profuse watery diarrhea.

ATR: Acid tolerance response, mechanism of microbial cells

that enables them to survive low-pH conditions. It is triggered by exposure of the cells to pH levels below 7.0 for specific periods of time.

Competitive Exclusion: Mechanism whereby harmless bacteria residing in the intestinal tract of humans and animals serve to exclude pathogens by taking up all available attachment sites on the intestinal mucosa.

Cytokines: Compounds produced by immune system cells such as PMNs. They serve a multitude of functions, including the signaling for specific T cells to help eliminate infected body cells. Examples are the interleukins and Tumor Necrosis Factor.

Endotoxins: Components of the outer membrane of gram-negative bacteria, which elicit a toxic effect on the host.

Exotoxic: Toxic substances produced by some pathogenic organisms, which are secreted by the cell.

Fecal-Oral Route: Route of transmission of foodborne pathogens, where organisms residing in feces of humans or animals contaminate the fingers, or drinking water, or utensils used for food preparation, or food itself. These then serve as vehicles to carry pathogens into the body through contact with the mouth.

Ileal Loop Test: Bioassay used to determine the presence of toxigenic bacterial cells. A rabbit is anesthetized, and its intestines are pulled out and ligated at various points to create sacs, or loops. The sample homogenate is injected into one of the ligated loops and the intestines are placed back into the abdominal cavity of the animal. After some time, the animal is sacrificed and the intestines examined to see if the ligated loop has increased in size. This is an indication that water diffused into the loop as a result of toxin being present in the sample.

Kupffer Cells: Macrophages residing in the liver.

LLO: Listeriolysin O, a compound produced by *Listeria monocytogenes*, which lyses red blood cells. It also enables the pathogen to survive inside of macrophages, and helps remove mucus from the surface of the intestinal mucosa. This enables the pathogen to adhere to the surface better. It is considered one of the main virulence factors of this organism.

Lysogen: A virus that integrates its genetic material into the chromosome of the host cell. The latter is not lysed by the virus, but rather continues to grow and divide, replicating the viral genes as it replicates its own chromosome.

Macropinocytosis: The uptake of foreign bodies (i.e. bacteria) by mammalian tissue cells, such as intestinal epithelial cells.

Microvilli: Epithelial cells lining the small intestine, which have protruding, "finger-like" appendages on the side facing the intestinal lumen. This helps facilitate nutrient uptake by increasing the surface area of the cell.

Multilocus Enzyme Electrophoresis: Technique used to examine differences among bacteria in a population. Since each individual enzyme produced by an organism represents a discrete and independent characteristic of that particular isolate, this technique can be used to differentiate among isolates of the same bacterial species. The method consists of subjecting cell extracts to electrophoresis in non-denaturing starch gels. Enzymes travel down the gel according to their electric charge, and are then reacted with their substrate. The position of the enzyme on the gel is thus determined by the appearance of a visible reaction product. Variations in the mobility of the enzyme down the gel reflects amino acid substitutions that change the charge of the protein. These changes are due to variations in the chromosomal genes encoding each enzyme. Thus, the pattern for a specific isolate represents a particular "multilocus genotype".

nvCJD: New variant Creutzfeldt-Jakob Disease, it is suspected to be caused by the prion that causes "mad cow" disease.

PMNs: Polymorphonuclear leukocytes, are white blood cells whose main function is to phagocytize foreign bodies. They circulate through the body via the lymphatic system. PMNs produce enzymes and oxygen radicals that help inactivate and kill the ingested intruder, and release compounds known as cytokines.

RpoS: Sigma factor that controls the induction of specific genes that help the cell survive environmental stressors, such as heat, starvation, and low pH conditions.

Septicemia: Infection of the circulatory system by virulent microorganism, which can then reach body tissues to begin new infections.

Sigma Factor: Protein subunit of the enzyme RNA polymerase which plays a role in translation of genetic material during synthesis of proteins. An organism can have several sigma subunits, with each one being responsible for translation of specific genes.

Stx: Shiga-like toxin, produced by specific strains of *Escherichia coli*, such as O157:H7.

Suckling Mouse Assay: Mice are inoculated intragastrically with a preparation of the suspected organism, and the animals are sacrificed after 4 hrs. The ratio of intestinal tract weight to body weight is measured. A ratio >0.83 is a positive test for the presence of toxin. This assay is carried out to detect heat-stable toxin produced by Enterotoxigenic *E. coli* strains, as well as other pathogens.

T Cell-Mediated Immunity: Mechanism of the immune system in which T cells are involved. Examples are T helper cells, T suppressor cells, Natural Killer (NK) cells, and Cytotoxic T Lymphocytes (CTLs). In general, T cells are involved in inducing the destruction of tissue cells that have been invaded (by pathogens such as bacteria and viruses) by NK and CTLs, and help regulate production of antibodies by B cells.

Virulon: Set of genes that control and code for specific virulence factors in bacteria.

Basic Knowledge Questions

1. List the five hurdles that a microorganism must overcome to cause disease.
2. What is the Sigma[38] factor?
3. List the sites of infection of the following pathogenic organisms:
   Helicobacter
   Hepatitis A
   Vibrio cholerae
   Trichinella spiralis
4. What is LLO and how does it help *Listeria monocytogenes* cause disease?
5. What are pathogenicity islands?
6. How do scientists think Enterohemorrhagic (EHEC) *E. coli* strains developed?
7. List and explain the three types of secretory systems of gram-negative bacteria.

Critical Thinking Questions

1. If you were to construct the ultimate foodborne pathogen, what virulence mechanisms would it have, and why?
2. Present at least two examples of how immune system cells are used by some pathogenic bacteria to further their ability to cause disease in the host.

CHAPTER 23

Staphylococcal Gastroenteritis

Learning Objectives

The reader will be able to:

1. Discuss coagulase-positive and coagulase-negative strains of *S. aureus* and their role in foodborne illness.
2. Discuss the distribution of *S. aureus* in nature and the types of foods where it can be found.
3. Discuss the growth requirements of this organism.
4. List and discuss the various toxin types of *S. aureus*, including their chemical and physical characteristics.
5. Discuss the conditions in which toxin production is optimal.
6. Discuss the disease and symptoms caused by intoxication with *S. aureus*.
7. Discuss strategies to prevent staphylococcal foodborne illness.

Chapter Outline

1. History.
2. Species of Concern in Foods.
   a) Coagulase-positive strains
   b) Coagulase-negative strains
3. Habitat and Distribution.
4. Incidence in Foods.
5. Nutritional Requirements for Growth.
6. Temperature Growth Range.
7. Effect of Salts and Other Chemicals.
8. Effect of pH, Water Activity, and Other Parameters.
   a) NaCl and pH
   b) pH, a_w, and temperature
   c) $NaNO_2$, Eh, pH, and temperature of growth
9. Staphylococcal Enterotoxins: Types and Incidence.
   a) Toxin types
   b) Chemical and physical properties
   c) Production
   d) Mode of action
10. The Gastroenteritis Syndrome.
11. Incidence and Vehicle Foods.
12. Ecology of *S. aureus* Growth.
13. Prevention of Staphylococcal and Other Food-Poisoning Syndromes.

Terms and Definitions

Coagulase: Enzyme produced by some staphylococci which causes clumping or clotting of blood plasma. Production of this enzyme was considered as an indicator of the ability of strains to produce enterotoxin. However, several coagulase-negative strains have been found to produce toxin.

De novo Protein Synthesis: Synthesis of proteins by synthesizing amino acids "from scratch," as compared with synthesis of proteins utilizing amino acids already present.

ELISA: Enzyme-linked immunosorbent assay, refers to a method used to detect the presence of specific organisms or toxins in a food or clinical sample. It is based on the use of antibodies specific for a particular antigen (microorganism or toxin). If the sample contains the antigen, the antibodies will bind to it, anchoring it to the surface of the microtiter plate. After washing any unbound antigen, a second set of antibodies specific for the antigen is then added, which are themselves linked to an enzyme. These will bind to the antigen, if present on the plate. Addition of substrate will result in an enzyme-catalyzed reaction, producing a color compound.

Emesis: Vomiting.

Enterotoxin: Toxin produced by a microorganism that affects the intestinal tract.

Gel Diffusion Assay: Also known as "Ouchterlony test," is a method used to confirm the presence of a toxin or other antigen in a sample. A layer of agar is poured into a dish or a glass plate. The gel is usually pre-mixed with a solution of antibodies specific for the antigen to be examined. After solidifying, holes 1–2 mm in diameter are punched into the gel slab to provide wells for the antigen samples. Antigen solution is added to a well and the plate is incubated. As the antigen diffuses into the gel, it forms a precipitate when it reacts with antibodies in the gel. The size of the precipitate ring around the well is indicative of the antigen concentration.

MHC: Major histocompatibility complex, refers to protein molecules present on the surface of animal cells. T cells of the immune system have receptors that recognize either foreign cells (containing MHC Class I molecules) or body cells (containing MHC Class II molecules). MHC class II molecules not only help T cells identify other body cells, but they also serve to aid in the communication between immune system cells, such as between macrophages and T cells.

TNAse: Thermonuclease (also known as heat-stable nuclease), refers to an enzyme produced by some staphylococci which digests DNA. It is relatively resistant to heating.

Basic Knowledge Questions

1. Explain what is meant by coagulase-positive or coagulase-negative, and list several species of staphylococci under each of these categories.
2. What effect does water activity, pH, and temperature have on growth and toxin-production of *S. aureus*?
3. Based on the types of conditions in which the organism grows, what types of food products are prone to harbor this organism?
4. List the toxin types produced by *S. aureus*, and describe the general properties of these, comparing them according to heat resistance and quantities produced.

5. List the symptoms and state the onset of illness caused by *S. aureus*.
6. What food products are more frequently associated with staphylococcal food poisoning?
7. List, in order, the five most common practices that yield to staphylococcal illness.

Critical Thinking Questions

1. How does pH and water activity affect production of specific toxin types by *S. aureus*?
2. What is the mode of action of staphylococcal toxins? What experiment(s) would you conduct to determine whether this is correct?

CHAPTER 24

Food Poisoning Caused by Gram-positive Sporeforming Bacteria

Learning Objectives

The reader will be able to:

1. Discuss the general characteristics of *Clostridium perfringens* and where it is found in nature.
2. Describe the toxins produced by *C. perfringens* and how they operate to cause disease.
3. Discuss the symptoms of the foodborne disease caused by *C. perfringens* and how this organism can be transmitted via foods.
4. Discuss how to prevent foodborne illness by this organism.
5. Describe the general characteristics of *Clostridium botulinum* and where it is found and distributed in nature.
6. Discuss the toxin types of *C. botulinum* and their mechanism of action.
7. Describe how certain food products and processing methods (i.e. sous vide) could increase the risk of botulism.
8. Define and describe infant botulism.
9. Discuss the characteristics of *Bacillus cereus*.
10. Describe the types of toxin produced by *B. cereus* and the disease syndromes they cause.

Chapter Outline

1. *Clostridium perfringens* Food Poisoning.
   a) Distribution of *C. perfringens*
   b) Characteristics of the organism
   c) The enterotoxin
      1) Toxin types
      2) Mode of action
   d) Vehicle foods and symptoms
   e) Prevention
2. Botulism.
   a) History
   b) Distribution of *C. botulinum*
   c) Growth of *C. botulinum* strains
   d) Ecology of *C. botulinum* growth
   e) Concerns for Sous Vide and related food products
   f) Nature of the botulinal neurotoxins
   g) The adult botulism syndrome: incidence and vehicle foods
   h) Infant botulism
3. *Bacillus cereus* Gastroenteritis.
   a) General characteristics
   b) *B. cereus* toxins
   c) Diarrheal syndrome
   d) Emetic syndrome

Terms and Definitions

Boticin E:	Bacteriocin produced by certain spoilage organisms that inhibits type E *Clostridium botulinum* cells.

Cereolysin: Hemolysin produced by *Bacillus cereus*. It is similar to perfringolysin but does not play a virulence role.

Cereulide: A water-insoluble peptide that is structurally similar to the antibiotic valinomycin. It is relatively heat-stable, and is produced by emetic strains of *B. cereus*.

Dialysis Sac: A method used to produce and separate spores of *C. perfringens* from vegetative cells of the organism. Generally, cells are inoculated into liquid medium lacking certain nutrients, and placed inside a dialysis sac with a pore size smaller than that of spores. The sac is placed inside a container with the same growth medium. During incubation, some of the cells will sporulate due to the lack of certain nutrients. The cells that do not sporulate will migrate across the dialysis membrane into the medium on the outside, leaving the spores inside the sac.

Hemolysin BL: Hemolytic peptide consisting of three parts (components B, L_1, and L_2), which causes lysis of red blood cells, cellular lysis, and necrosis of epithelial cells, among other effects.

Lactic Antagonism: The inhibition of certain pathogenic bacteria by lactic acid bacteria, such as *Lactobacillus* and *Lactococcus*.

Mouse LD: Mouse lethal dose, it refers to the amount of a toxin required to kill a particular number of mice. It can be expressed as mouse LD_{50}, which refers to the concentration of toxin required to kill 50% of a certain number of mice.

Perfringolysin: Hemolytic compound produced by *C. perfringens* which plays a role as a virulence factor in causing disease.

Sous Vide: Processing treatment whereby foods are cooked at relatively low temperatures under vacuum in order to retain flavors and improve the overall organoleptic quality. It is believed that this process may provide the right conditions such that growth of *C. botulinum* cells would take place preferentially, since their competitors are destroyed during cooking.

Tenesmus: Urge to defecate with inability to empty the bowels at the same time.

Terminal/Sub-Terminal Spores: Terminal spores are those that develop at one end of a vegetative cell, making the sporulated organism appear as a club-shaped body under the microscope. Sub-terminal spores are those that develop between the center and the end of the vegetative cell.

Trypsin: Enzyme that digests some proteins at specific amino acid sites. It causes the breakage of certain chemical bonds in botulinal progenitor toxin, activating it.

Basic Knowledge Questions

1. List the toxins produced by *C. perfringens*.
2. What are the factors that affect heat-resistance of spores?
3. What is the relationship between enterotoxin production and sporulation of *C. perfringens*?
4. List at least 3 recommendations that should be followed to avoid foodborne illness outbreaks due to *C. perfringens*.
5. What are the various groupings of toxin-producing *C. botulinum* strains, and what are some of their characteristics? Include in your answer relative heat resistance, growth temperature and habitat.

6. Discuss the mechanism of action of botulinal toxin and the symptoms it elicits.
7. Describe the diarrheal and emetic syndromes caused by *B. cereus*, including the toxins responsible and symptoms experienced.

Critical Thinking Questions

1. In testing for the presence of botulinal toxin in a food sample, the mouse bioassay is employed, in which mice are injected intraperitoneally to induce symptoms. You are asked to use this assay to determine whether a particular food contains type A toxin. Design an experiment, including the controls that you would use, to test the sample for the presence of this type of toxin.
2. Speculate as to why the diarrheal syndrome caused by *B. cereus* does not require as many cells to be present as with the emetic syndrome.

CHAPTER 25

Foodborne Listeriosis

Learning Objectives

The reader will be able to:

1. Discuss the taxonomical classification of *Listeria* species.
2. Describe the nutritional requirements of *Listeria* species, including the effect that various parameters have on growth of these organisms (i.e. temperature, pH, a_w).
3. Discuss how *Listeria* species are distributed in the environment, and the types of foods that have been found to harbor these organisms, especially *L. monocytogenes*.
4. Describe the thermal properties of *L. monocytogenes*, particularly within the context of milk pasteurization, and including the effect of heat shock on acquired thermotolerance.
5. Describe the mechanisms of virulence of *L. monocytogenes*, including how it invades and survives within phagocytic cells.
6. Discuss the methods used to determine infectious dose of this pathogen.
7. Describe the symptoms of listeriosis and the disease syndromes it causes.
8. Discuss the mechanisms of resistance to listeriosis in humans, and the regulatory limits that have been imposed on ready-to-eat foods in various countries to prevent outbreaks of this disease.

Chapter Outline

1. Taxonomy of *Listeria*.
   a) General characteristics
   b) Serotypes
   c) Subspecies typing
2. Growth.
   a) Nutritional requirements
   b) Effect of pH
   c) Combined effect of pH and NaCl
   d) Effect of temperature
   e) Effect of a_w
3. Distribution.
   a) The environment
   b) Foods and humans
   c) Prevalence
4. Thermal Properties.
   a) Dairy products
   b) Nondairy products
   c) Effect of sublethal heating on thermotolerance
5. Virulence Properties.
   a) Listeriolysin O and Ivanolysin O
   b) Intracellular invasion
   c) Monocytosis-producing activity
   d) Sphingomyelinase
6. Animal Models and Infectious Dose.
7. Incidence and Nature of the Listeriosis Syndromes.

a) Incidence
b) Source of pathogens
c) Syndromes
8. Resistance to Listeriosis.
9. Persistence of *L. monocytogenes* in Foods.
10. Regulatory Status of *L. monocytogenes* in Foods.

Terms and Definitions

ActA: Surface protein of *L. monocytogenes*, which induces the production of actin filaments that help propel the cell through the cytoplasm of phagocytes en route to the membrane. Once there, the bacterium can be transmitted to another phagocyte through the formation of vacuoles that transport the organism across the membrane of one cell to that of another.

Anton Test: Assay used to confirm the presence of invasion-type of microorganisms, such as *L. monocytogenes*. A culture of the suspected organism is added to the eye of a rabbit or guinea pig. A positive reaction is depicted by the development of conjunctivitis.

Bacteremia: The presence of bacteria in the bloodstream.

Beta Hemolysis: Lysis of red blood cells in agar medium, seen as a complete zone of clearing surrounding the colonies of bacteria responsible for it. *L. monocytogenes* is an example.

CAMP Test: Christie, Atkins, Munch-Petersen, refers to a biochemical test that is used to confirm the identity of *Listeria monocytogenes*. The suspected culture is streaked on an agar plate containing sheep blood. An inoculum of *Rhodococcus equi* or *Staphylococcus aureus* is added by streaking the culture perpendicularly to the listerial inoculum. After incubation, *L. monocytogenes* will cause hemolysis of the red blood cells in the agar, with the reaction being magnified at the points where the two organisms intersect.

Filopodium: Structure similar to pseudopods which helps in the transfer of *L. monocytogenes* from one phagocytic cell to another across the membrane.

FIM: Factor-increasing monocytopoiesis, refers to a compound produced by macrophages after ingestion of *L. monocytogenes* cells. This factor is secreted and stimulates production of more macrophages by the immune system.

Gamma-Interferon: Cytokine compound produced by T cells, it elicits the production of other cytokines (i.e. Interleukin-1). This induces Natural Killer cells (a type of T cell) to destroy macrophages infected with organisms like *L. monocytogenes*.

Granulocytes: Polymorphonuclear leukocytes, or white blood cells of the immune system. Examples are neutrophils, basophils, and eosinophils. They contain granules within the cytoplasm, which are used to digest invading microorganisms during phagocytosis.

HTST: High-temperature, short-time pasteurization, typically carried out in milk at 71.7°C for 15 seconds.

ILO: Ivanolysin O, refers to a compound produced by *Listeria ivanovii*. It is similar to listeriolysin O, and causes a complete zone of hemolysis immediately around red blood cells.

Internalins: Proteins found on the surface of *L. monocytogenes*, required for the organism to invade epithelial cells (Internalin A) and mouse hepatocytes in culture (Internalin B).

LAL Assay: Limulus amoebocyte lysate assay is a method used to detect LPS of gram-negative bacteria. Aliquots of food homogenates are added to LAL reagent, which is a component found in amoebocytes of the horseshoe crab. It reacts with LPS of the bacteria.

LLO: Listeriolysin O, refers to the main virulence factor of *L. monocytogenes*. It has a molecular weight of 60,000 daltons and is responsible for the hemolytic activity of the organism. It causes death of phagocytes that have ingested the bacteria.

LTLT: Low-temperature, long-time pasteurization, also known as vat-pasteurization, it is typically carried out at 62.8°C for 30 minutes.

Lipoteichoic Acid: Component of the cell wall peptidoglycan of some bacterial cells.

Meningitis: Inflammation of the meninges, or the membrane surrounding the brain and spinal chord.

Monocytes: Phagocytic cells of the immune system, also knows as macrophages. In general, the term "monocyte" is used to refer to a specific type of macrophage: those that are in the circulatory system. The term "macrophages" refers to those that are fixed to specific tissues (i.e. liver).

Nude Mice: Mice that are deficient in their ability to produce T cells. They are born without hair.

Phagolysosomes: Membrane-bound cavities inside the cytoplasm of phagocytes. They are formed by the fusion of a phagosome (the cavity formed when a phagocyte engulfs and internalizes an invading microorganism into its cytoplasm) and a lysosome (cytoplasmic vacuole containing lytic enzymes that help in digestion of invading bacteria). Most engulfed microorganisms are destroyed inside phagolysosomes, with the notable exception of *L. monocytogenes*.

Phospholipases: Enzymes produced by *L. monocytogenes* that serve to partially digest the membrane of phagocytic cells to achieve vacuole formation in order to transport the organism to another cell.

Septicemia: Infection of the bloodstream, such as caused by bacteria.

Sphingomyelinase: Enzyme produced by some *Listeria* species, responsible for incomplete hemolysis observed near the *Rhodococcus equi* colonies in the CAMP test.

"Zero Tolerance": Policy that allows no cells of a particular pathogen to be detected in 25- or 50-g samples of foods. For example, the U.S. has a "zero tolerance" regarding *L. monocytogenes* in ready-to-eat foods.

Basic Knowledge Questions

1. List the five species of *Listeria*.
2. Describe the habitats of *Listeria* species, and their prevalence in foods.

3. How is *Listeria* disseminated in the environment?
4. What are LLO and ILO?
5. Describe the effects of infection with this pathogen on both healthy and immuno-compromised patients.
6. List some of the sources of infection of this organism in nature, and in the food processing plant.
7. What is the current policy regarding *L. monocytogenes* in the U.S. and other countries?

Critical Thinking Questions

1. What is the role of T cells in the initiation of infection by *L. monocytogenes*, as well as in resisting infection by this organism?
2. Various studies have been conducted to prove or disprove the notion that *L. monocytogenes* can survive pasteurization. Review the section of this chapter in which this topic is discussed, and present a theory as to why there is such a disparity. What type of experiment would you conduct to settle the issue one way or another?

CHAPTER 26

Foodborne Gastroenteritis Caused by *Salmonella* and *Shigella*

Learning Objectives

The reader will be able to:

1. Describe the taxonomy of salmonellae and how it has changed over the years.
2. Describe the methods used to serotype the organism, and list the major types that are involved in foodborne illness.
3. Discuss the distribution of salmonellae in nature, including food products.
4. Describe the general characteristics of salmonellae, including growth and their susceptibility to processing treatments.
5. Discuss the virulence factors of salmonellae and their role in causing disease.
6. Discuss the types of foods in which salmonellae are found and which have been linked to infection by these organisms, as well as what can be done to prevent it.
7. Define "competitive exclusion" and discuss how this concept is used to prevent colonization of salmonellae in live animals.
8. Describe the main characteristics of shigellae and their role in foodborne illness.
9. Discuss the symptoms of shigellosis and the virulence factors that lead to this infection.

Chapter Outline

1. Salmonellosis.
   a) Taxonomy
   b) Serotyping
   c) Distribution
      1) Animals
      2) Animal feeds
      3) Food products
   d) Growth and destruction of salmonellae
   e) The *Salmonella* food-poisoning syndrome
   f) Incidence and vehicle in foods
   g) Prevention and control of salmonellosis
   h) Competitive exclusion to reduce salmonellae carriage in poultry
2. Shigellosis.
   a) General characteristics of *Shigella* species
   b) Foodborne cases
   c) Virulence properties

Terms and Definitions

Alpha Amylase Test: Assay used to determine whether liquid egg has been treated at the correct temperature to achieve pasteurization. With such a treatment, alpha-amylase is inactivated.

Nurmi Concept: This concept was first developed to explain how colonization of animal tissue by some bacteria could be prevented as a result of competition by other microorganisms living in the same environment. It is also known as "competitive exclusion."

S. senftenberg 775W: Serotype of salmonellae recognized as the most heat resistant. It requires longer heat treatment than other serotypes to eliminate it from foods.

Transovarial: Traveling from the ovaries (of the hen) into the developing embryo inside an egg. It is considered one way in which eggs could be contaminated with salmonellae.

Basic Knowledge Questions

1. List and describe the three groups into which salmonellae are placed for epidemiological purposes.
2. Describe the system used to serotype salmonellae, based on O and H antigens.
3. List the types of products where salmonellae are typically found.
4. Name the top 3 *Salmonella* serotypes in terms of their being isolated most frequently from clinical samples.
5. Describe "competitive exclusion" as applied to poultry flocks.
6. List the three species of shigellae responsible for outbreaks of foodborne illness.
7. What is the most common factor involved in the incidence of shigellosis worldwide?

Critical Thinking Question

1. Provide a theory as to why *Salmonella* Enteritidis infections are on the rise. What routes of transmission could account for its prevalence in eggs and in poultry?

CHAPTER 27

Foodborne Gastroenteritis Caused by *Escherichia coli*

Learning Objectives

The reader will be able to:

1. Describe the classification scheme used for *Escherichia coli* serotypes.
2. Describe the general characteristics of Enteroaggregative *E. coli*, including its prevalence and its role in disease.
3. Describe the general characteristics of Enterohemorrhagic *E. coli*.
4. Discuss the toxins produced by enterohemorrhagic *E. coli*, including the circumstances in which it is optimally produced and how certain conditions can affect it.
5. Describe the prevalence of Enterohemorrhagic *E. coli* in foods and its origin in animals.
6. Describe the symptoms of disease caused by Enterohemorrhagic *E. coli*.
7. Describe Enteroinvasive and Enteropathogenic *E. coli* serotypes, including the disease syndromes they cause.
8. Describe the general characteristics of Enterotoxigenic *E. coli*, and the toxins produced by this organism.
9. Discuss the disease syndrome caused by Enterotoxigenic *E. coli*.

Chapter Outline

1. Serological Classification.
2. The Recognized Virulence Groups.
   - a) Enteroaggregative *E. coli*
   - b) Enterohemorrhagic *E. coli*
     - 1) The toxins
     - 2) Growth and Stx production
     - 3) Effect of environmental and physical agents
     - 4) Prevalence in foods
     - 5) Prevalence in dairy cattle
     - 6) Human disease syndromes/prevalence
   - c) Enteroinvasive *E. coli*
   - d) Enteropathogenic *E. coli*
   - e) Enterotoxigenic *E. coli*
     - 1) The enterotoxins
3. Prevention.
4. Traveler's Diarrhea.

Terms and Definitions

A/E: Attachment/effacement, refers to the mechanism by which several types of pathogenic *E. coli*, such as Enterohemorrhagic strains, colonize the intestinal tract. The bacteria attach to tissue cells, damaging them and causing an effacement of the lining of the intestinal tract. This mechanism has been linked to the possession of a particular gene by the bacterium, designated "*eae*A."

CFA: Colonization factor antigen, refers to fimbrial elements produced by Enterotoxigenic *E. coli* strains, which help them colonize intestinal tissue cells.

EAST1: Heat-stable enterotoxin produced by some Enteroaggregative *E. coli* strains.

Enterobacteriaceae: Family of bacteria to which several genera of important foodborne pathogens belong, such as *Salmonella*, *Escherichia*, *Shigella*, *Yersinia*, and others.

Enteroinvasive: Refers to the property of a pathogenic organism in terms of its ability to invade intestinal tissue cells.

Esps: Enteropathogenic-*E. coli*-secreted-proteins, refers to proteins produced by these organisms, which block phagocytosis and cause phosphorylation of certain surface receptors of intestinal tissue cells. The latter facilitates the intimate attachment of the *E. coli* cells to the microvilli of the intestinal cells.

Hemolytic Anemia: One of the manifestations of hemolytic uremic syndrome, in which red blood cells lyse, resulting in a decrease in their number. This develops in some children as a result of infection with Enterohemorrhagic *E. coli* strains.

G Protein: Protein found inside intestinal tissue cells. Some bacterial toxins, such as that produced by Enterotoxigenic *E. coli* strains, cause ribosylation of the G protein, which itself activates the adenylate cyclase system. This activation results in production of cAMP which disrupts the transport of ions by the intestinal cells, causing efflux of water into the intestinal lumen.

HUS: Hemolytic uremic syndrome, refers to the syndrome developed in children due to infection generally with Enterohemorrhagic *E. coli* strains. It results in kidney damage which often necessitates dialysis and can ultimately end in death of the patient.

Intimin: Protein produced by several pathogenic strains of *E. coli*, such as Enteropathogenic and Enterohemorrhagic strains. It allows for the intimate binding of the bacteria to the microvilli of intestinal tissue cells, setting the stage for either their invasion or effacement.

LT: Heat-labile toxin, refers to a toxin produced by Enterotoxigenic *E. coli* strains, made up of an A and a B subunit. The A subunit stimulates the adenylate cyclase system while subunit B helps the toxin bind to intestinal tissue cells. The toxin is destroyed by heating at 60°C for 30 min.

OMP: Outer membrane protein, such as that in the outer membrane of Enteroaggregative strains of *E. coli*, believed to be responsible for adherence to tissue cells.

ST: Heat-stable toxins produced by Enterotoxigenic strains of *E. coli*. They withstand heating at 100°C for 15 minutes. These toxins stimulate the guanylate cyclase system.

Sereny Test: Assay performed to determine whether a bacterial strain is able to invade tissue cells. Drops of a culture are added to the eye of a guinea pig, with development of conjunctivitis deemed as a positive for infectivity.

Stx Toxins: Also known as Shiga-like toxins, verotoxins, and SLT-1 and SLT-2 toxins. There are two major Stx toxins, Stx1 and Stx2, as well as some variants. The toxins bind to gangliosides, and become internalized into the cell. The mechanism of action of the toxins is to inhibit protein synthesis by causing the release of an adenine residue from the rRNA of ribosomes.

Thrombo-Cytopenia: Damage to thrombocytes resulting in clot formation due to infection with Enterohemorrhagic *E. coli* strains. This occurs in the elderly, and can result in embolism, or stroke.

Basic Knowledge Questions

1. Define the following: EaggEC, EHEC, EIEC, and ETEC.
2. What two toxins are produced by EHEC, and what is their mode of action?
3. How do pH and temperature affect production of toxins by EHEC?
4. Describe the disease conditions caused by EHEC.
5. How do Esps help EIEC cause disease?
6. What are the modes of action of the two toxins produced by ETEC?
7. What organisms are responsible for causing traveler's diarrhea?

Critical Thinking Questions

1. What could be some reasons why the rate of incidence of EHEC in Northern Ireland, England, Wales, and Scotland are so much lower than in the U.S.?
2. Why do you think that calves shed *E. coli* O157:H7 longer and are usually infected with this organism more frequently than adult cattle?

CHAPTER 28

Foodborne Gastroenteritis Caused by *Vibrio*, *Yersinia*, and *Campylobacter* Species

Learning Objectives

The reader will be able to:

1. Describe the various *Vibrio* species that are responsible for foodborne illness, in terms of growth conditions and prevalence in nature.
2. Describe the virulence factors of *Vibrio parahaemolyticus* and the disease syndrome it causes.
3. List and describe other *Vibrio* species and explain how they cause foodborne illness.
4. Describe *Yersinia enterocolitica* in terms of its growth requirements and distribution in nature.
5. List the serovars and biovars associated with yersiniosis and their incidence in foods.
6. Describe the virulence factors of *Y. enterocolitica* and the disease it causes.
7. Describe the general characteristics of *Campylobacter jejuni* and its distribution in nature.
8. Explain the virulence factors of *C. jejuni*, specifically the toxins it produces, and describe the disease syndrome it causes.

Chapter Outline

1. Vibriosis (*Vibrio parahaemolyticus*).
   a) Species
   b) Growth conditions
   c) Virulence properties
   d) Gastroenteritis syndrome and vehicle foods
2. Other Vibrios.
   a) *Vibrio cholerae*
   b) *Vibrio vulnificus*
   c) *Vibrio alginolyticus* and *V. hollisae*
3. Yersiniosis (*Yersinia enterocolitica*).
   a) Growth requirements
   b) Distribution
   c) Serovars and biovars
   d) Virulence factors
   e) Incidence of *Y. enterocolitica* in foods
   f) Gastroenteritis syndrome and incidence
4. Campylobacteriosis (*Campylobacter jejuni*).
   a) General characteristics
   b) Distribution
   c) Virulence properties
   d) Enteritis syndrome and prevalence
5. Prevention.

Terms and Definitions

Cholecystitis:	Inflammation of the gall bladder, usually due to infection with a microorganism, such as *Y. enterocolitica*.
CJT:	*Campylobacter jejuni* toxin, refers to a heat-labile endotoxin produced by this organ-

ism. It is composed of one A subunit surrounded by five B subunits. Its mode of action is through activation of the adenylate cyclase system, much like *V. cholerae* toxin and Enterotoxigenic *E. coli* LT toxin.

Guillain-Barré Syndrome: Chronic condition caused by prior infection with some organisms, most commonly *Campylobacter jejuni*. It manifests itself as a temporary paralysis of the facial muscles.

El Tor: Biotype of serovar O group 1 *V. cholerae*. It was first recognized as unique due to its ability to agglutinate chicken red blood cells.

Erythema Nodosum: Chronic skin disease, marked by the appearance of tender pink nodules, usually caused by infection with a microorganism such as *Y. enterocolitica*.

Inaba: Specific serotype within the El Tor biotype of *V. cholerae*. It is serologically distinct from the Ogawa serotype.

Isoelectric Point: pH of a molecule at which the net molecular charge is zero. At this point, migration through a pH gradient stops.

Kanagawa Reaction: Assay used to determine the presence of virulent strains of *V. parahaemolyticus*. It is based on the development of a beta-hemolysis reaction on Wagatsuma's agar. The hemolysin responsible for this is thermostable.

NAGs: Non-agglutinating vibrios, refers to strains of *V. cholerae* that do not react with antibodies specific for serovar O group 1 vibrios, and thus have been placed into their own.

NAG-ST: Toxin produced by non-agglutinating vibrios. It is similar to toxins produced by *Citrobacter freundii* and *Yersinia enterocolitica*.

Ogawa: Specific serotype within the El Tor biotype of *V. cholerae*. It is serologically distinct from the Inaba serotype.

OmpU: Outer membrane protein which has been shown to play a role in adherence of *V. cholerae* strains to intestinal tract cells, a necessary step for invasion.

Penner: Serotyping scheme used to differentiate *Campylobacter jejuni* isolates, based on differences due to heat-stable antigens on the cell surface. Approximately 60 serotypes have been identified according to this system.

Peritonitis: Inflammation of the peritoneal cavity due to fluid accumulation as a result of infection with certain bacteria, such as *Y. enterocolitica*.

Pseudo-Appendicitis: Condition attributed to infection with *Y. enterocolitica* in which the patient suffers from similar symptoms to appendicitis. Cases of yersiniosis have sometimes led to appendectomies by physicians unfamiliar with this pathogen.

R Plasmids: Plasmid DNA that contains gene(s) for resistance to specific antimicrobial agents. It is transferred between species of the same genus, but can be transferred across genera as well.

Reactive Arthritis: Condition caused by infection with some foodborne pathogens, notably *Y. enterocolitica* and some salmonellae serotypes. It can develop 7 days after infection, and is marked by inflammation of the joints, especially wrists and knees. It is believed to result from stimulation of T helper cells to induce production of cytotoxic T cells by the engulfing of bacteria by macrophages. The CTLs then produce certain cytokines which continue stimulation of the T helper cells, resulting in fluid accumulation.

TDH: Thermostable direct hemolysin, responsible for the Kanagawa reaction. It is produced by most virulent strains of *V. parahaemolyticus*.

TRH: Thermostable-related hemolysin, hemolysin present in some strains of *V. parahaemolyticus*. In addition to TDH, it is an indicator of virulence for these organisms.

YOPs: Yersinia outer-membrane proteins, refers to a class of about 20 proteins produced by strains of *Y. enterocolitica*, which may help the bacteria adhere to intestinal tissue cells, and even escape phagocytosis. Proteins are encoded by genes carried in plasmids.

Basic Knowledge Questions

1. Define the Kanagawa reaction and explain its role in the virulence of *Vibrio* species.
2. List the various biovars and serotypes of *V. cholerae.*
3. List the minor species of *Vibrio* that have been connected with foodborne illness (other than *V. parahaemolyticus* and *V. cholerae*), and provide a short description of the types of outbreaks they have caused.
4. Describe the major characteristics of *Y. enterocolitica* and list the serotypes that have been identified.
5. What are the virulence factors of *Y. enterocolitica*?
6. Describe the major characteristics of *C. jejuni* and how it differs from most other foodborne pathogens.
7. What is the principal virulence factor of *C. jejuni* and what is its mechanism of action?

Critical Thinking Questions

1. Provide an explanation for the fact that despite the major differences among these organisms, *V. cholerae, C. jejuni*, and Enterotoxigenic *E. coli* all possess a toxin with the same mode of action. Consider environmental as well as other factors in your answer.
2. In spite of the fact that *C. jejuni* is a fastidious organism, able to grow only under very strict conditions in the laboratory, it is considered the leading cause of diarrhea in developing countries. How can this be explained?

CHAPTER 29

Foodborne Animal Parasites

Learning Objectives

The reader will be able to:

1. List the main differences between bacteria and animal parasites.
2. Describe the protozoa, and provide a list of organisms as examples.
3. Describe the protozoan diseases giardiasis, amebiasis, toxoplasmosis, sarcocystosis, cryptosporiodiosis, and cyclosporiasis, in terms of organisms responsible, foods associated that serve as vehicles of infection, disease symptoms, and the treatment that should be applied.
4. Describe the main characteristics of flatworms and roundworms.
5. Describe the diseases caused by flatworms, such as fascioliasis, fasciolopsiasis, paragonimiasis, clonorchiasis, diphyllobothriasis, and taeniasis, including the organisms responsible, the foods associated with these infections, disease symptoms, and treatment and prevention measures.
6. Describe the causative agent of trichinosis, the disease symptoms it causes, the mode of infection, and how this can be prevented.
7. Describe the causative agent of anisakiasis, the disease symptoms it causes, the mode of infection, and how this can be prevented.

Chapter Outline

1. Differences from Other Microorganisms.
2. Protozoa.
   a) Taxonomy
   b) Giardiasis
      1) Environmental distribution
      2) Syndrome, diagnosis, and treatment
      3) Incidence in foods and foodborne cases
   c) Amebiasis
      1) General characteristics
      2) Syndrome, diagnosis, and treatment
   d) Toxoplasmosis
      1) General characteristics
      2) Symptoms, diagnosis, and treatment
      3) Distribution of *T. gondii*
      4) Food-associated cases
      5) Control
   e) Sarcocystosis
   f) Cryptosporiodiosis
      1) General characteristics
      2) Symptoms, diagnosis, and treatment
      3) Waterborne and foodborne outbreaks
   g) Cyclosporiasis
      1) General characteristics
      2) Prevalence and outbreaks
3. Flatworms.
   a) Classification
   b) Fascioliasis
      1) General characteristics
      2) Symptoms, diagnosis, and treatment
   c) Fasciolopsiasis
   d) Paragonimiasis

1) General characteristics
2) Symptoms, diagnosis, and treatment
e) Clonorchiasis
1) General characteristics
2) Symptoms, diagnosis, treatment, and prevention
f) Diphyllobothriasis
1) General characteristics
2) Prevalence
3) Symptoms, diagnosis, and treatment
4) Prevention
g) Cysticercosis/Taeniasis
1) General characteristics
2) Symptoms, diagnosis, and treatment
3) Prevention
4. Roundworms.
a) Classification
b) Trichinosis
1) General characteristics
2) Prevalence
3) Symptoms and treatment
4) Diagnosis
5) Prevention and control
6) Microwave cooking
c) Anisakiasis
1) General characteristics
2) Symptoms, diagnosis, and treatment
3) Prevalence and distribution
4) Prevention

Terms and Definitions

Amebiasis: Disease caused by infection with *Entamoeba histolytica*, also knows as amoebic dysentery. Onset of symptoms is 2–4 weeks, with loose stools containing mucus and blood lasting for several months, or even years. These develop into severe abdominal pain, fever, vomiting, and back pain, with weight loss a typical outcome.

Anisakiasis: Disease caused by infection with the roundworms *Anisakis simplex* and *Pseudoterranova decipiens*. Onset of symptoms is very fast, detected as early as within 6 hours after ingestion. Stomach pain, nausea, vomiting, with fever and bloody stool possible in some cases. Peritonitis can result if the worms are able to penetrate the gut wall.

Bradyzoites: Dormant forms of tachyzoites, develop after cluster of tachyzoites become encircled by a protective wall inside muscle and other tissues. They do not multiply but can persist in an infected person for life. When the immune system of the victim is compromised, bradyzoites will break, resulting in the emergence of tachyzoites.

Cercariae: Larval stage of some trematodes such as *F. hepatica*, having the shape of a tadpole with a tail-like appendage. This form is produced by a redia, and are able to swim and attach to grasses.

Clonorchiasis: Disease caused by infection with the trematode *Clonorchis sinensis*. Onset of symptoms is similar to that of other fluke-associated illness (approximately 1 month). An infected person may be asymptomatic, but in some cases liver damage takes place, which can lead to liver cancer.

Cryptosporiodiosis: Disease caused by infection with *Cryptosporidium parvum*. Onset of symptoms is approximately 2 weeks, and these include voluminous watery diarrhea, impaired respiration, dehydration, abdominal discomfort, which develop into weight loss. These can last about 4 days.

Cyclosporiasis: Disease caused by infection with *Cyclospora cayetanensis*. Onset and symptoms are similar to cryptosporiodiosis, but can last 6 weeks or more.

Cysticercosis: Disease caused by infection with the flatworms *Taenia saginata* or *Taenia solium*. Symptoms are usually not experienced at all. However, in 60% of infected person, cysticerci develop in the central nervous system resulting in symptoms like epileptic seizures. In 5% of infected persons, cysticerci form in the muscle yielding to muscle enlargement. In 3% of cases they form in the eye.

Definitive Host: The animal in which an adult parasite carries out its sexual cycle and multiplies.

Diphyllobothriasis: Disease caused by infection with the parasite *Diphyllobothrium latum*. Onset of disease is approximately one month, with symptoms including abdominal cramps, vomiting, dizziness, followed by loss of appetite yielding weight loss.

Fascioliasis: Disease caused by infection with the trematode *Fasciola hepatica*. It is also known as parasitic biliary cirrhosis, or liver rot. Onset is approximately 30 days, consisting of fever, fatigue, loss of appetite, and pain in the liver region of the abdomen.

Fasciolopsiasis: Disease caused by the parasite *Fasciolopsis buski*. Onset is 1–2 months, with symptoms including violent diarrhea, abdominal pain, weakness, and weight loss.

Giardiasis: Disease caused by infection with the protozoan *Giardia lamblia*, carried through meat or water contaminated with animal feces harboring this organism. Onset of symptoms is 7–13 days, with cramps, nausea, severe diarrhea, vomiting, and flatulence that result in weight loss. Can last up to 3 months and is considered highly contagious.

Intermediate Host: The animal in which a juvenile or larval form of a parasite develops and matures.

Metacercaria: Encysted form of cercariae of trematodes such as *F. hepatica*. This is the form that is typically ingested by a person. Once inside the body, it excysts, traveling to the liver where it matures in the bile ducts.

Nosocomial: Refers to an infection acquired in hospital.

Oocysts: Term used to describe the encysted, inactive, form of most protozoans. Oocysts will often develop or excysts to active forms, which cause disease.

Paragonimiasis: Disease caused by the parasites *Paragonimus westermani* and *P. kellicotti*. Onset is 2–3 months, with symptoms including severe chronic coughing and sharp chest pain.

Proglottids: Segments of the body of a tapeworm, such as *T. saginata*, containing both male and female reproductive organs.

Rediae: Larvae produced within the sporocysts of certain trematodes, such as *F. hepatica*. They develop into cercariae.

Sarcosystosis: Disease caused by *Sarcocystis hominis* or *S. suihominis*. Onset of symptoms is relatively short (3–48 hours), and include nausea, stomach cramps, and diarrhea.

Scolex: Head of a flatworm, such as *T. saginata*, either in the larval or adult stage.

Sporozoites: Life forms that result from germination of *T. gondii* oocysts. These pass through the intestinal walls and are able to reach the circulatory system. From there, they are transported to muscle tissue and other parts of the body, where they develop into tachyzoites.

Strobila: The chain of segment (proglottids) forming the body of a tapeworm, such as *T. saginata*.

Tachyzoites: Term used to refer to the rapidly-multiplying form of sporozoites of *T. gondii* and similar parasites after reaching muscle tissue and organs. Tachyzoites will form clusters surrounded by a protective wall, developing into a bradyzoite.

Toxoplasmosis: Disease caused by the protozoan *Toxoplasma gondii*. It is transmitted through contact with infected farm animals, as well as cats. Most people do not develop symptoms, but the immunocompromised will suffer from symptoms that mimic infectious mononucleosis. These include rash, headache, fever, severe muscle pain, and swelling of the lymph nodes.

Trematode: A class of parasite also known as flukes. They have a well-developed alimentary canal and suckers on their outer surface for adhesion to tissues.

Trichinosis: Disease caused by infection with the roundworm *Trichinella spiralis*. Onset of symptoms is approximately 2 weeks, and these include severe muscle pain, facial swelling, diarrhea, and vomiting, with death resulting in some cases due to heart failure.

Trophozoite: Flagellated form of *G. lamblia* after maturation of cysts takes place. They penetrate the intestines through a falling-leaf type of motility. The same term is used to describe the ameba stage of *E. histolytica*, which are also motile.

Zoonotic: Refers to an infection acquired through direct contact with a vertebrate animal.

Basic Knowledge Questions

1. What is the difference between a definitive host and an intermediate host?
2. Describe the life cycle of *Giardia lamblia* that yields to disease in humans.
3. What are trophozoites, and what role do they play in causing parasitic disease? Include in your answer the types and names of parasites which occur in the form of trophozoites.
4. Compare and contrast the organisms *Cryptosporidium parvum* and *Cyclospora cayetanensis* in terms of general characteristics, how transmitted, and the diseases they cause.
5. Compare the diseases caused by the trematodes discussed in this chapter.
6. Describe the life-cycle of *Taenia*, including the various forms that are involved.
7. How is the risk of trichinosis prevented/minimized?

Critical Thinking Questions

1. Should we trust the use of microwaving to ensure the safety of fresh meats from the parasite *Trichinella*? Why or why not? Be complete in your answer, providing known evidence for your point of view.
2. What measures could/should be implemented if we were to significantly reduce the incidence of parasitic diseases in humans due to consumption of contaminated raw foods?

CHAPTER 30

Mycotoxins

Learning Objectives

The reader will be able to:

1. Describe aflatoxins in terms of general structures and the conditions under which they are produced by certain molds.
2. Describe the prevalence of aflatoxins in foods, and the mode of action responsible for disease.
3. Describe alternaria toxins, citrinin, ochratoxins, patulin, penicillic acid, and sterigmatocystin, in terms of the organisms that produce them, their structure, the types of foods where found, and their mode of action.
4. Describe fumonisins, including the conditions under which they are produced, the organisms that produce them, the foods where they are found, and the disease conditions they cause.
5. Describe sambutoxin and zearalenone toxins, including the organisms that produce them, the types of foods where found, and the disease conditions they cause.

Chapter Outline

1. General Definitions.
2. Aflatoxins.
   a) Structures
   b) Requirements for growth and toxin production
   c) Production and occurrence in foods
   d) Relative toxicity and mode of action
   e) Degradation
3. Alternaria Toxins.
4. Citrinin.
5. Ochratoxins.
6. Patulin.
7. Penicillic Acid.
8. Sterigmatocystin.
9. Fumonisins.
   a) Toxin types
   b) Growth and production
   c) Prevalence in corn and feeds
   d) Physical/chemical properties of FB_1 and FB_2
   e) Pathology
10. Sambutoxin.
11. Zearalenone.
    a) Structure
    b) Control and production

Terms and Definitions

Aflatoxins: Toxic compounds produced by some *Aspergillus* mold species. They are secondary metabolites that can cause abdominal pain, jaundice, and vomiting if ingested. If dose consumed is high enough, it can result in liver necrosis. Consumption of low doses for extended periods of time may lead to liver cancer.

Alternariol: Mycotoxin produced by *Alternaria* species, not proven to be harmful to humans.

Ames Test: Assay used to determine the mutagenicity of chemical compounds. It is based on the genetic change of a *Salmonella* Typhimurium culture, from one that requires certain nutrients to grow, to one that is able to grow without the nutrient being supplied.

Citrinin: Mycotoxin produced by *Penicillium citrinum* and *P. citreoviridae*. It is heat-stable and causes tubular damage in kidneys of swine as ochratoxins, yet it is somewhat less toxic than ochratoxin.

Cyclopiazonic Acid: Compound produced by some *A. flavus* strains, thought to contribute to the toxicity of aflatoxin.

Fumonisins: Mycotoxins produced by *Fusarium* species. There are at least seven fumonisins, with FB_1 being associated with esophageal cancer in humans.

Hyperestrogenism: Symptoms mimic those caused by having very high levels of estrogen. Females get enlarged and edematous genitals resulting in abortions, and males undergo a feminizing effect, having enlarged mammary glands and atrophied testes. It is caused by consumption of zearalenone mycotoxin.

Ochratoxin: Mycotoxin produced by *Aspergillus ochraceous* and some *Penicillium* species. Type A is most toxic and is heat-stable to 169°C. It affects the liver and kidneys, causing necrosis of the tubules in rats, dogs, and swine. May cause a renal disorder in humans called Balkan Endemic Nephropathy.

Patulin: Mycotoxin produced by *Penicillium expansum*, carcinogenic to mice. It is acid-stable and heat-stable. It is suspected to be carcinogenic to humans, but it has not yet been proven.

Primary Metabolites: Compounds produced by microorganisms that are essential for growth, and exerted during active growth.

Sambutoxin: Mycotoxin produced by *Fusarium sambucinum* and *F. oxysporum*. It causes hemorrhaging in the stomach and intestines of rats, and it is toxic to chick embryos.

Secondary Metabolites: Compounds produced at or near the end of exponential growth phase. They have no apparent usefulness to the organism in terms of growth or metabolic capacity.

Trichothecene: Mycotoxin produced by *Fusarium sporotrichioides*, also known as T_2 toxin. It is responsible for alimentary toxic aleukia, a condition that causes a burning sensation in the mouth, followed by acute gastroenteritis. This is followed by destruction of the bone marrow, development of leukemia, anemia, yielding to total atrophy of the bone marrow and lung hemorrhaging, ultimately resulting in death.

Zearalenone: Mycotoxin produced by *Fusarium graminearum*, also known as F-2 toxin. It causes estrogenic syndrome in swine. It is heat-stable.

Basic Knowledge Questions

1. What is the difference between primary and secondary metabolites?
2. List the various types of aflatoxins, and compare them in terms of toxicity and other characteristics.
3. Draw the chemical structures of citrinin, ochratoxin, patulin, and penicillic acid.
4. List and describe the various fumonisins, and list the organisms that produce them.
5. Describe the disease syndromes caused by consumption of trichothecenes.
6. What is sambutoxin?
7. What is hyperestrogenism and what organism is responsible for causing it?

Critical Thinking Question

1. There are several mycotoxins that are known to affect only animals. Design an experiment that would prove/disprove the role of fumonisins in causing esophageal cancer in humans.

CHAPTER 31

Viruses and Some Other Proven and Suspected Foodborne Biohazards

Learning Objectives

The reader will be able to:

1. Define viruses and describe the methods used to detect them in foods.
2. Discuss the prevalence of viruses in nature and in foods, and how they can be eliminated from foods.
3. Describe the main characteristics of Hepatitis A virus, of Norwalk and related viruses, and of rotaviruses.
4. Discuss the role of bacteria in Scombroid food poisoning.
5. Discuss the major characteristics and the illnesses caused *by Aeromonas, Plesiomonas, Bacteroides fragilis, Erysipelothrix rhusiopathiae, Klebsiella pneumoniae*, and *Sreptococcus iniae.*
6. Define the term "prion" and describe how these structures may be responsible for "mad cow disease" in cattle, and related illnesses in humans.
7. Describe the toxigenic phytoplanktons that cause paralytic shellfish poisoning, ciguatera poisoning, and poisoning caused by domoic acid.
8. Describe the main characteristics of *Pfisteria* and its suspected role in causing foodborne illness.

Chapter Outline

1. Viruses.
   a) Detection methods
   b) Incidence in foods and the environment
   c) Destruction in foods
   d) Hepatitis A virus
   e) Norwalk and related viruses
   f) Rotaviruses
2. Bacteria and Prions.
   a) Histamine-associated (Scombroid) poisoning
   b) *Aeromonas*
   c) *Plesiomonas*
   d) *Bacteroides fragilis*
   e) *Erysipelothrix rhusiopathiae*
   f) *Klebsiella pneumoniae*
   g) *Streptococcus iniae*
3. Prion Diseases.
   a) Mad cow disease
   b) Cruetzfeld-Jakob Disease
4. Toxigenic Phytoplanktons.
   a) Paralytic shellfish poisoning
   b) Ciguatera poisoning
   c) Domoic acid
   d) *Pfisteria piscicida*

Terms and Definitions:

ASP: Amnesic shellfish poisoning, caused by domoic acid produced by the diatom *Pseudonitschia pungens*.

BSE: Bovine spongiform encephalopathy, also known as "mad cow disease," refers to an illness in cattle belonging to the family of diseases called "tranmissible spongiform encephalopathies." It is caused by a prion, and results in uncoordination and brain damage.

CJD: Creutzfeldt-Jakob disease, refers to an illness belonging to the family of diseases called "transmissible spongiform encephalopathies." It is caused by a prion infecting some individuals. It is thought that a harmless configuration of the prion is naturally found in people. Certain unknown conditions are thought to cause a conformational change in the prion to a form that causes holes in brain tissue, ultimately resulting in death of the infected individual. It is also suspected that a new variant of the prion, termed nvCJD, is found in cattle and that it can cause CJD by infection from animals to humans.

Ciguatera: Food poisoning due to consumption of tropical herbivorous fish which themselves have ingested the dinoflagellate *Gambierdiscus toxicus*. Onset of disease is within 5 hours, resulting in diarrhea, nausea, vomiting, and abdominal pain. Several hours later, neurological symptoms develop, such as hot/cold inversion, metallic taste, dizziness, blurred vision, and temporary blindness.

Diatom: Single-celled alga with silicon-based walls.

Dinoflagellate: Member of the algae family, it is aquatic in nature, photosynthetic, and some produce toxin. Consumption of particular dinoflagellates by certain fish result in various seafood toxin syndromes.

Domoic Acid: Toxin produced by the diatom *Pseudonitschia pungens*. It causes a disease from consumption of mussels infected with the dinoflagellate. It excites the brain cells until they die from overstimulation, resulting in amnesia.

Mad Cow Disease: Common term used for the degenerative disease called bovine spongiform encephalopathy. The disease is thought to be caused by prions infecting the cattle.

Paresthesia: Tingling, especially around the mouth.

Prion: Small proteins that, upon assuming a specific conformational shape, can cause brain damage in animals and humans. They aggregate into fibrils, causing nerve cell degeneration.

RT-PCR: Reverse-transcription polymerase chain reaction, refers to a method of identification of foodborne viruses when only a few plaque forming units are available (viral particles). Foodborne viruses contain RNA instead of DNA as their genetic material. In order to identify them, the samples containing viral RNA must be reacted with the enzyme reverse transcriptase in order to synthesize DNA from the RNA molecules. At this point, the polymerase chain reaction assay can be applied in order to augment the number of copies of viral DNA. With an increased number of DNA molecules, gel electrophoresis can then be applied to detect the viral DNA and thus the particular organism in the sample.

Saxitoxin: Toxin produced by the dinoflagellate *Gonyaulax catenella*. It causes paralytic shellfish poisoning due to consumption of mussels, clams, and other shellfish that have ingested the dinoflagellate. May cause death due to respiratory paralysis within a few hours.

Scombroid Poisoning: Disease caused by consumption of scombroid fish such as mahi-mahi, sardines, and anchovies. These contain high levels of free histidine in their muscles. This compound serves as a substrate for the enzyme histidine decarboxylase produced by various spoilage bacteria. Rash, edema, nausea, vomiting, diarrhea, abdominal cramps, headache, palpitations, tingling, flushing, and burning sensation in the mouth are some of the symptoms due to the presence of histamine produced from the break down of histidine.

SRSVs: Small round structural viruses, of which the Norwalk virus is a member. These viruses belong to the family *Caliciviridae*.

Basic Knowledge Questions

1. What is the incidence of viruses in foods in the U.S.?
2. List and describe the three groups of Norwalk and Norwalk-like viruses.
3. What is histamine food poisoning caused by?
4. What organisms does *Aeromonas* resemble and in what way?
5. What are prions and how do they cause disease?
6. Describe some of the disease syndromes caused by toxins produced by dinoflagellates.
7. What is *Pfiesteria piscicida*?

Critical Thinking Question

1. Describe the evidence that supports or contradicts the theory that prions harbored in cattle are responsible for CJD in humans.